Molecular Ecology of Rhizosphere Microorganisms

Edited by
F. O'Gara, D. N. Dowling, B. Boesten

VCH

Distribution:

VCH, P. O. Box 101161, D-69451 Weinheim, Federal Republic of Germany

Switzerland: VCH, P. O. Box, CH-4020 Basel, Switzerland

United Kingdom and Ireland: VCH, 8 Wellington Court, Cambridge CB1 1HZ, United Kingdom

USA and Canada: VCH, 220 East 23rd Street, New York, NY 10010–4606, USA

Japan: VCH, Eikow Building, 10-9 Hongo 1-chome, Bunkyo-ku, Tokyo 113, Japan

ISBN 3-527-30052-X

Molecular Ecology of Rhizosphere Microorganisms

Biotechnology and the release of GMOs

Edited by
F. O'Gara, D. N. Dowling, B. Boesten

VCH

Weinheim · New York
Basel · Cambridge · Tokyo

Editors:
Prof. Fergal O'Gara
Dr. David Dowling
Ir. Bert Boesten
BioMerit
Microbiology Department
University College
Cork
Ireland

This book was carefully produced. Nevertheless, authors, editors and publisher do not warrant the information contained therein to be free of errors. Readers are advised to keep in mind that statements, data, illustrations, procedural details or other items may inadvertently be inaccurate.

Published jointly by
VCH Verlagsgesellschaft mbH, Weinheim (Federal Republic of Germany)
VCH Publishers Inc., New York, NY (USA)

Editorial Director: Dr. Hans-Joachim Kraus
Production Manager: Dipl.-Wirt.-Ing. (FH) H.-J. Schmitt

Library of Congress Card No.: applied for

British Library Cataloguing-in-Publication Data:
A catalogue record for this book
is available from the British Library

Die Deutsche Bibliothek – CIP-Einheitsaufnahme:
Molecular ecology of rhizosphere microorganisms :
biotechnology and the release of GMOs / ed. by F. O'Gara ... –
Weinheim ; New York ; Basel ; Cambridge ; Tokyo : VCH, 1994
ISBN 3-527-30052-X
NE: O'Gara, Fergal [Hrsg.]

Printed on acid-free and chlorine-free (TCF) paper

Composition: Filmsatz Unger & Sommer, D-69469 Weinheim
Printing: Strauss Offsetdruck, D-69509 Mörlenbach
Bookbinding: Großbuchbinderei J. Schäffer, D-67269 Grünstadt
Printed in the Federal Republic of Germany.

Preface

Techniques in biotechnology and information science are revolutionising our ability to understand how microbes interact with their environment. The emerging field of molecular microbial ecology will allow technologists, researchers and regulators to monitor, model and predict with increased accuracy the outcome of a range of microbial applications of key biotechnological importance including; health-care (clinical microbiology), agribusiness (biocontrol and food microbiology) and environmental microbiology (bioremediation). This book is a collection of papers presented by the speakers and tutors at an EC sponsored BRIDGE Advanced Workshop in Biotechnology on "The Molecular Ecology of Rhizosphere Bacteria" held in Cork (22nd March – 2nd April 1993). The workshop consisted of an international research forum, co-organised by BIOMERIT (EC Comett II programme) and a practical component directed towards young researchers.

The genetic and molecular techniques that can be applied to the study of the ecology of rhizosphere microorganisms are as numerous and diverse as the microbes themselves. Experimental methods developed for unraveling the molecular complexity of the cell are being directed to the study of rhizosphere ecosystems and the integration of molecular methods with classical methods is expanding our understanding of rhizosphere microbial ecology. Biotechnology has been a major impetus in applying new methods in rhizosphere ecology. The availability for release of Genetically Modified Microorganisms (GMOs) has stimulated research programmes to evaluate their potential impact in the rhizosphere.

The chapters cover different areas of rhizosphere microbiology. They provide an overview of the current concepts and bottlenecks in our understanding of the molecular basis of rhizosphere microbial ecology and the impact that GMOs may have on this ecosystem. The contents of this book represent the synthesis of the authors contributions to the workshop which we hope will go some way towards defining a molecular basis to understanding rhizosphere microbial ecosystems.

We would like to thank the authors for their interest and committment to the workshop and the members of the plant-microbiology group (UCC) who contributed to its success. The encouragement and support of Dr A. Leonard and Dr I. Economidis of the European Commission (DG12) were appreciated. Finally, we would like to thank Sheila Kelleher, Joan Buckley and Mary Cotter for their secretarial assistance.

Fergal O'Gara
David N. Dowling
Bert Boesten
Editors

List of Contributors

David M. Weller and Linda S. Thomashow
USDA-Agricultural Research Service,
Root Disease and Biological Control Research Unit,
Pullman, Washington 99164-6430,
U.S.A.

John A. McInroy and Joseph W. Kloepper
Department of Plant Pathology,
Biological Control Institute,
Alabama Agricultural Experiment Station,
Auburn University,
Auburn, Alabama 36849, USA.

Maarten H. Ryder[1,2], Clive E. Pankhurst[1,2], Albert D. Rovira[1], Raymond L. Correll[3] and Kathy M. Ophel Keller
[1] Cooperative Research Centre for Soil and Land Management,
[2] CSIRO Division of Soils and
[3] CSIRO Biometrics Unit
Glen Osmond SA 5064
Australia

J.M. Lynch, F. A.A. M de Leij,[1] J.M. Whipps[1] and M.J. Bailey[2]
School of Biological Sciences,
University of Surrey,
Guildford, Surrey, GU2 5XH, UK
[1] Horticulture Research International,
Littlehampton, West Sussex,
BN17 6LP, UK
[2] NERC Institute of Virology and Environmental Microbiology,
Mansfield Road, Oxford, OX1 3SR,
UK

David N. Dowling, Bert Boesten, Paul R. Gill, Jr. and Fergal O'Gara
Department of Microbiology,
University College,
Cork,
Ireland

Christophe Voisard [1,2,+], Carolee T. Bull[2,+], Christoph Keel[1], Jacques Laville[2], Monika Maurhofer[1], Ursula Schnider[2,+], Geneviève Défago[1] and Dieter Haas[2*]
[1] Department of Plant Sciences/Phytomedicine and
[2] Department of Microbiology,
Eidgenössische Technische Hochschule,
CH-8092 Zürich,
Switzerland.
[+] Present address: Swiss Meteorological Institute, CH-8044 Zürich,
Switzerland.
[*] Laboratoire de Biologie Microbienne,
Université de Lausanne,
CH-1015 Lausanne,
Switzerland.

Víctor de Lorenzo
Centro de Investigaciones Biológicas-CSIC;
Velázquez 144, Madrid 28006, Spain.

Hans-Volker Tichy and Reinhard Simon
TÜV Südwest,
Fachgruppe Biologische Sicherheit,
Robert-Bunsen-Straße 1,
D-79108 Freiburg, Germany.

J.E. Cooper and A.J. Bjourson
Plant Pathology Research Division
Department of Agriculture for Northern Ireland
Newforge Lane, Belfast BT9 5PX,
Northern Ireland

Pascal Simonet, Sylvie Nazaret and Philippe Normand.
Laboratoire d'Ecologie Microbienne du Sol,
URA CNRS 1450,
Université Claude Bernard Lyon I,
43 Bd du 11 Novembre 1918,
69622 Villeurbanne, France.

Peter Marsh and Elizabeth M. H. Wellington
Department of Biological Sciences,
University of Warwick, Coventry, UK.

J.D. van Elsas and E. Smit
Institute for Soil Fertility Research,
P.O. Box 7060, 6700 GW Wageningen,
the Netherlands.

Marco P. Nuti[1,2], Andrea Squartini[1] and Alessio Giacomini[1]
[1] Dipartimento di Biotecnologie agrarie,
Università di Padova,
via Gradenigo 6, 35121 Padova (Italy)
[2] CRIBI Biotechnology Centre,
Università di Padova, Complesso "A. Vallisneri",
via Trieste 75, 35121 Padova, (Italy)

Content

1 Current Challenges in Introducing Beneficial Microorganisms into the Rhizosphere

David M. Weller and Linda S. Thomashow

1.1 Introduction and Definitions

Since 1904, when Lorenz Hiltner introduced the term rhizosphere, much has been learned about its biology, microbiology, and ultrastructure. The rhizosphere is the narrow zone of soil surrounding the root that is subject to the influence of the root. Intense microbial activity and larger microbial populations occur in this zone as compared to the bulk soil, in response to the release from roots of large amounts of organic matter (50–100 mg/g of root), in the form of exudates, lysates, and mucilages. As much as 18% of carbon assimilated as photosynthate can be released from roots. Organic compounds lost from roots include sugars, amino acids, organic acids, fatty acids, nucleotides, vitamins and enzymes. Since the rhizosphere is rich in exudates, the microbial population can reach up to 1×10^9 cells per cm^3, 10–100 times larger than the population in the bulk soil. Rhizosphere microorganisms include bacteria, viruses, fungi, arthropods, mites, amoebae, and flagellates. The rhizosphere extends away from the root for 1–2 mm, but some organisms may be stimulated up to 5 mm away.

The rhizoplane refers to the actual surface of the root; however, as a root ages, cortical cells undergo autolysis (a genetically controlled trait) and the boundary between the rhizoplane and the rhizosphere becomes blurred. The root cortex becomes colonized by microorganisms such that only the tissues of the stele remain alive. Thus, a part of the root becomes an extension of the rhizosphere known as the endorhizosphere. The use of the term endorhizosphere recently has been questioned (1). Interestingly, despite the intense microbial activity in the rhizosphere, only about 7–15% of the actual root surface (rhizoplane) is covered with microorganisms. They are clumped into microcolonies in sites where nutrients are most abundant. These include grooves between epidermal cells, root hairs, lesions and sites where lateral roots break through cortical cells. The rhizosphere is a dynamic environment, and rhizosphere interactions have substantial impact on plant growth and development (2,3).

Rhizobacteria are plant-associated bacteria that are able to colonize and persist on roots (4). Rhizobacteria are subdivided into beneficial, deleterious and neutral groups on the basis of their effects on plant growth. Studies during the late 1970s and early 1980s at the University of California, Berkeley demonstrated that certain fluorescent *Pseudomonas* strains, termed plant growth-promoting rhizobacteria (PGPR), could improve the growth of potato and sugar beet when applied to seeds or seed pieces (5,6). The results of these studies along with public concerns about the adverse affects of chemical pesticides helped to catalyze a resurgence of research worldwide on bacterial inoculants to control pathogens and improve plant growth. The term PGPR is now applied to a wide spectrum of strains that have, in common, the ability to promote the growth of plants following inoculation onto seeds or subterranean plant parts (4). Growth promotion can occur through direct stimulation of the plant either by increasing the supply of mineral nutrients, such as phosphorous and nitrogen, or by the production of phytohormones (7). PGPR also improve growth indirectly through the suppression of major and minor soilborne pathogens (8,9). Major pathogens produce the well-known root or vascular diseases with obvious symptoms. Minor pathogens are parasites or saprophytes that damage mainly juvenile tissue such as root hairs and tips and cortical cells, and produce symptoms that are not obvious (10). Schippers et al. (11) distinguished the parasitizing minor pathogens from the nonparasitizing deleterious rhizosphere microorganisms (DRMO). DRMO include deleterious rhizobacteria (12) and deleterious fungi. PGPR have been identified from many genera besides *Pseudomonas* these include *Agrobacterium, Alcaligenes, Arthrobacter, Azospirillum, Azotobacter, Bacillus, Bradyrhizobium, Enterobacter, Erwinia, Flavobacteria, Hafnia, Klebsiella, Rhizobium, Serratia* and *Xanthomonas.*

Several definitions of root colonization by rhizobacteria have been proposed (4,13,9). In this chapter it is defined as the process whereby bacteria introduced on seeds or vegetatively propagated plant parts become distributed along roots growing in raw soil, multiply, and survive for several weeks in the presence of indigenous soil microflora. Root colonization includes colonization of the root surface, the inside of the root and/or the rhizosphere. Rhizosphere competence describes the relative root-colonizing ability of a rhizobacterium. The rhizosphere competence of a strain can be quantified by determining the population size it attains on a root, the length or number of roots colonized and/or the length of time the bacteria survive (9).

In spite of the "biological buffering" that generally limits introducing microorganisms into soil (14), PGPR can be established successfully because they are highly adapted to the rhizosphere. Further, the application of large populations allows rapid occupation of preferred sites where exudates are greatest.

1.2 Relationship of Root Colonization to Biocontrol and Growth Promotion

It is generally accepted that PGPR must become positioned on or in the root or in the rhizosphere to promote growth (7,15,16,17). Rhizobacteria growing in or near infection courts on roots, as well as in channels in the rhizosphere that provide physical access to the root, are ideally positioned to limit the establishment and spread of pathogens. Several studies have demonstrated that PGPR suppress populations of root pathogens (18,19,20). For example, Xu and Gross (21) applied *Pseudomonas putida* W4P63 to potatoes and monitored its population and that of its target *Erwinia carotovora* on roots in the field. The populations of W4P63 ranged between 10^4 and 10^5 colony forming units per g root, while the population of *E. carotovora* on the same root was only 10% of that on roots without W4P63. However, the threshold populations required to achieve pathogen suppression or growth promotion have not been well defined for most PGPR. Bull et al. (22) studied the relationship between the population size (ranging from 10^1-10^5 per cm root) of *P. fluorescens* 2–79 established on wheat roots via seed treatments, and the number of lesions caused by the take-all pathogen, *Gaeumannomyces graminis* var. *tritici*. Linear regression analysis demonstrated an inverse relationship between the population size of 2–79 and the number of lesions, thus indicating that as colonization increased, take-all control improved.

Another important question is the duration that populations of PGPR must be maintained in order to close the "window of vulnerability" to infection by pathogens. For some diseases, such as *Pythium* seed rot and damping-off, protection may be required for only hours to days, whereas, for diseases like take-all, protection will need to last weeks to months. In general, the smaller the "window of vulnerability" the greater are the chances of successful biocontrol.

1.3 The Process of Colonization

How are PGPR translocated from sources of inoculum on seeds or seedpieces to critical sites along the root and in the rhizosphere? Howie et al. (23) hypothesized a two phase process. In Phase I, bacteria on the seed attach to the emerging root tip and are passively transported into the soil. During root growth some cells remain associated with the tip while others are left behind on older portions of the root and in the rhizosphere. Bacteria may be dislodged as the root extends through the soil or become adsorbed to soil particles (24). In Phase II, bacteria deposited along the root multiply and form microcolonies in nutrient-rich microsites, compete with indigenous microflora and avoid displacement. Any bacterium applied to the seed can be transported into the soil with the roots but only those that are rhizosphere competent will maintain or increase their population (25). The concept of Phase I and II is meant to define two stages in the life history of introduced PGPR, rather than a strict temporal sequence of events because both phases occur simultaneously on different parts of the roots. Although percolating water is not required for root colonization (22,23,24), Liddell and Parke (15) demonstrated that it can make a major contribution to the long distance transport of introduced PGPR. They applied *P. fluorescens* PRA25 to pea seed and monitored root colonization in a Plainfield sand incubated at 24–26 °C. In the absence of infiltrating water PRA25 was restricted to root segments 0–4 cm below the seed at matric potentials of −1, −6 and −10 kPa.

In contrast, PRA25 was recovered from root segments up to 13–14 cm below the seed following the application of 27.2 mm of water. Percolating water can wash bacteria down the root directly from the inoculum source, recharge root tips that have outgrown the bacteria with inoculum from older portions of the root (24) and redistribute bacteria from established microcolonies into new microsites. The relative importance of root tip and water transport is highly dependent on the bacterial strain, host plant, soil type, temperature and amount of water. Undoubtedly, in the field the two processes are integrated.

1.4 Effect of Biotic and Abiotic Factors

The distribution, multiplication and survival of introduced PGPR are profoundly affected by biotic and abiotic factors. Howie et al. (23) studied the effect of soil matric potential on colonization of wheat roots by seed-applied *P. fluorescens* 2–79 in the absence of percolating water. The largest populations developed on roots at −30 Kpa in one soil and at −70 Kpa in two other soils. The range from −30 Kpa to −70 Kpa is probably where oxygen availability and turgor potential of the cells and/or nutrient availability were optimal for bacterial cell growth. Strain 2–79 was transported from seeds onto roots even at −400 Kpa.

The optimal temperature for growth of many PGPR in vitro is above 25 °C, but root colonization is generally greatest below 20 °C. For example, colonization of potato roots by two fluorescent *Pseudomonas* strains was best at 12 °C or 18 °C (16). Gutterson et al. (26) reported that *P. fluorescens* Hv37a colonized cotton roots efficiently at 16 °C and 20 °C, but populations were reduced by up to 100 fold at 24 °C. Seong et al. (27) reported that *P. fluorescens* ANP15 colonized roots better at 18 °C than at 30 °C. *Pseudomonas fluorescens* 88W1 survived better in raw soil at 5 °C and 15 °C as compared to 25 °C (28). Better root colonization at lower temperatures probably reflects the fact that microbial activity (and thus competition) in the soil declines with temperature. Further, slower root growth at lower temperatures may facilitate more effective transport of the bacteria from the inoculum source to the roots.

Although PGPR grow best in vitro at neutral pH or above, colonization is better at lower pH, possibly because of lower competition. For example, colonization of wheat roots by *P. fluorescens* 2–79 was greater at rhizosphere pH 6–6.5 than at 7.0 or above (29).

The biological composition of the rhizosphere will dramatically influence root colonization. It is well understood that in the rhizoplane and rhizosphere nutrient availability rather than space is the primary determinant of microbial population size (30). Thus, introduction of PGPR does not result in a change in the total rhizosphere population, but a shift in the composition of the microflora such that introduced bacteria preempt establishment of the normal indigenous strains. Thus, root colonization will be greater in sterile or pasteurized soils than in raw soil because there is less competition, antibiosis and predation from the indigenous microflora. In contrast, as microbial activity increases in the soil, through inputs of nutrients, the level of colonization by introduced PGPR is reduced. Stephens et al. (31) reported that bacteriophage were responsible for at least part of the decline in the population of *P. fluorescens* B2/6 in the rhizosphere of sugar beet following introduction as a seed treatment. Protozoa and bacteria are also potential predators of introduced rhizobacteria (32,33).

Both the type and quantity of root exudates upon which rhizosphere microflora depend are under environmental and genetic control. Thus, the composition of the indigenous rhizosphere microflora, as well as the population size of introduced PGPR varies among plant species (34) and varieties of the same crop species (35,36,37). For example, when *P. fluorescens* 2-79 was applied as a seed treatment, its population size on the roots of the wheat variety Wampum was 100-fold greater than on the variety Brevor (38).

Pathogens that are targets of PGPR can influence PGPR populations either positively or negatively. For example, the population of 2-79 was larger on roots infected with *G. g. tritici* than on healthy roots (39), probably as a result of bacterial proliferation in lesions that are rich in nutrients. Similarly, root infection by *R. solani* resulted in significantly larger populations of both *P. fluorescens* 2-79 and Q72a-80 than were present on healthy roots (40). This is not surprising since *Rhizoctonia* also causes deep cancerous root lesions that increase the flow of nutrients. In contrast, the population of 2-79 was significantly smaller on roots infected with *Pythium irregulare, P. aristosporum*, or *P. ultimum* var. *sporangiiferum* than on noninfected roots. Interestingly, the effect of *Pythium* was strain-specific; the population size of Q72a-80 was reduced only in the presence of *P. irregulare*. Application of metalaxyl (selectively inhibits *Pythium* spp.) to a soil naturally infested with *Pythium* spp. resulted in significantly larger populations of 2-79 or Q72a-80 on roots as compared to roots from soil not treated with the chemical (40).

1.5 Bacterial Traits Contributing to Rhizosphere Competence

In that root colonization is a multistage process, undoubtedly many bacterial traits and genes are involved. The importance of each trait may differ amongst PGPR. Bacterial adherence to roots is probably one of the early steps in Phase I of root colonization, and currently is of considerable research interest. Several bacterial cell surface properties have potential for involvement. Adhesion of PGPR to roots may be either non-specific resulting from electrostatic forces (41), or involve specific recognition between the surfaces. For example, the ability of *P. putida* strain Corvallis to bind to and colonize bean roots and suppress cucumber wilt, caused by *Fusarium oxysporum* f. sp. *cucumerinum*, was correlated with agglutinability of the bacteria by a root surface glycoprotein termed an agglutinin (42,43). Buell and Anderson (44) characterized a locus, *agg*A, from *P. putida* that encodes a predicted 50.5 kDa protein, required for agglutinability and adherence. The predicted amino acid sequence revealed a signal peptide cleavage signal consistent with export of the putative protein from the cytoplasm, but no similarity to sequences in several databases. The distribution of the locus among plant-associated bacteria was limited to those expressing the agglutination phenotype. van Peer et al. (45) isolated *Pseudomonas* spp. from the surface and interior (endorhizosphere) of tomato roots and found that for endorhizosphere isolates, there was a significant correlation between agglutinability with a tomato root agglutinin and colonization of the endorhizosphere. However, no such correlation was found for root surface isolates. Chao et al. (46) demonstrated that a greater percentage of bacteria isolated from the pea rhizosphere as compared to the bulk soil were agglutinated by pea root agglutinin.

Pseudomonas fluorescens OE28.3, from the rhizosphere of wheat, produces a 32.1 kDa major outer membrane protein, a root adhesin, that adsorbs strongly and selectively to the roots of barley, maize and sunflower seedlings (47,48). The locus encoding the adhesin was characterized, and the deduced amino acid sequence showed strong homology with the amino- and carboxyterminal ends of porin F from *P. aeruginosa* and *P. syringae* (49). Porin F forms water-filled channels through the outer membrane and is an important structural protein for maintenance of cell shape and growth on media of low osmolarity (50).

Several different exopolysaccharides are involved in the attachment of *Agrobacterium tumefaciens* to plant cells (51,52,53) and in the nodulation of legumes by *Rhizobium* (54,55). In *A. tumefaciens,* mutations in either of two chromosomal virulence loci, *chv*A and *chv*B, resulted in impairment of attachment and avirulence. Interestingly, Waelkens et al., (56) reported that DNA homology to the *chv* genes was found in *Azospirillum brasilense* and *A. lipoferum*, PGPR that also attach to root surfaces (57). Further, cosmid clones from a library of *A. brasilense* DNA complemented *Rhizobium meliloti* mutants that were deficient in production of succinoglycan, the major acidic exopolysaccharide required for nodulation (58). These studies suggest that the early phases in the interaction between *Azospirillum* and roots may have similarities to those that occur with *Agrobacterium* and *Rhizobium*.

In some PGPR, fimbriae (pili) may function in adherence of cells to roots. One example is their mediation of adhesion of N_2-fixing strains of *Klebsiella* and *Enterobacter* to roots of grasses and cereals (59,60,61,62). Vesper (63) reported a positive correlation between the presence of fimbriae and the attachment of 2-79 to corn roots. Fimbriae are reported to play a role in the attachment of *Bradyrhizobium japonicum* and *Rhizobium trifolii* to soybean roots (64,65).

The contribution of flagella to the colonization process apparently depends on the PGPR strain, plant species, and type of soil. The moisture status of the soil is particularly important because in soil drier than −50 Kpa, water films probably are too thin and water-filled pores too small and discontinuous to allow flagella-mediated movement. Howie et al. (23) reported that colonization of wheat roots by flagelladeficient mutants of *P. fluorescens* R7z-80R, Rla-80R and R4a-80R was equivalent to that of their respective parental strains in two different soils and at matric potentials favorable (−20 Kpa) and unfavorable (−200 Kpa) for motility. Similarly, *P. putida* RW3 and its nonflagellated Tn*5* mutant applied as seed treatments developed similar populations on soybean roots (66). In contrast, each of four nonmotile Tn*5* mutants of *P. fluorescens* WCS374 applied to 1-cm-long roots of potato stem cuttings developed significantly smaller populations at a depth of 8 cm than the wild-type (67).

The importance of the production of antibiotics and other secondary metabolites to the biocontrol activity of many PGPR strains has been conclusively demonstrated (68). Recently, Mazzola et al. (69) demonstrated that phenazine (Phz) antibiotics also contribute positively to the persistence of *P. fluorescens* 2–79 and *P. chloroaphis* 30–84 in soil habitats. Strains 2-79, 30–84, phenazine-deficient mutants (Phz^-) and mutants complemented to Phz^+, individually were introduced into raw soil with or without *G. g. tritici*. Up to five cycles of wheat were sown, each lasting twenty days from planting to harvest. At the end of each cycle, shoots were severed and the soil and roots were removed, mixed, repotted and again sown to wheat. Populations of the Phz^- mutants declined significantly more rapidly than populations of their respective parental or genetically restored Phz^+ strains in both rhizosphere and bulk soil. The differences between Phz^+ and Phz^- strains appeared more rapidly in the absence than presence of *G. g. tritici*. Populations of Phz^- and respective Phz^+ strains remained similar when the studies were conducted in steam-pasteurized soil (reduced populations of soil microflora), suggesting that phenazine production contributes to competitiveness against indigenous microorganisms.

Several other bacterial traits may contribute to rhizosphere competence including chemotaxis toward seed or root exudates (70,71), ability to utilize root exudates and secretions, especially complex carbohydrates (72,73), rapid growth in the rhizosphere (13) and tolerance to dry soils and low osmotic potential (9,16).

1.6 Population Dynamics of PGPR in the Field

The rhizosphere competence of introduced PGPR and their spatial-temporal colonization patterns are evaluated best in field studies where the full, natural component of soil microflora and fauna are present (24). However, because such studies are labor-intensive and time-consuming colonization studies usually are conducted under controlled conditions. Those field studies that have been conducted have demonstrated that PGPR can become widely distributed along the root system. Bahme and Schroth (24), in the most elegant field study yet conducted, determined the spatial-temporal colonization patterns of *P. fluorescens* A1-B at all stages of plant development on below-ground parts of field-grown potatoes in a silty clay-loam and a sandy clay-loam. Bacteria applied at 10^8 cfu per seedpiece, eventually were isolated from root segments (up to 36–40 cm away from their origin at the stem), from progeny tubers, and from underground portions of shoots. Populations were greatest on plant parts nearest inoculated seed pieces. Mean population densities on roots (prior to irrigation) and progeny tubers were significantly larger in the sandy loam as compared to the silty clay-loam. Spatial distribution patterns and population densities on roots were substantially altered after irrigation.

P. fluorescens 2–79 has been used as a model organism to study the fate of introduced pseudomonads in the field on wheat roots. Bacteria applied to the seed

(10^8 cfu/seed) initially became distributed along the length of seminal roots with populations greatest on sections of root closest to the seed (approximately 10^6 cfu/cm root). There was a significant linear decline in the population along the axis from the seed to the tip with the population doubling every 15–85 hrs. (74). Populations of 2–79 on subcrown internodes initially were greater than 10^6 cfu per 0.1 g tissue. Crown roots became colonized by inoculum located at the base of the tillers and on the subcrown internode. The pattern of colonization of 2–79 paralleled closely the growth of *G. g. tritici* on wheat, which initially infects the seminal roots and spreads to the crown of the plant via the subcrown internode (39). Regardless of the strain applied, the population dynamics of introduced pseudomonads on the total root system followed a similar pattern on spring and winter wheat and on wheat grown in different soils and in different locations. Large populations initially became established on the roots and then the population size gradually declined over the growing season. For example, on wheat grown from seed treated with 10^8 cfu of 2–79 per seed and sown in October, by 18 days after planting 2–79 was present at 10^6 cfu/0.1 g root with adhering soil. However, at 245 days after planting the population declined by three orders of magnitude. Populations of 2–79 remained significantly higher on roots infected with *G. g. tritici* than on healthy roots.

1.7 Release of Genetically Engineered Rhizobacteria

In 1987, Kluepfel et al. (75) released the first genetically engineered rhizobacteria into a wheat field at Clemson University's Research and Education Center near Blackville, South Carolina. The *lac*ZY genes (76) were inserted into the chromosome of *P. aureofaciens* PS3732RN (rifampin and nalidixic acid resistant), using a Tn*7* vector derivative, generating the recombinant PS3732RNL11, which utilized lactose as a sole carbon source and produced blue colonies on media amended with the chromogenic dye X-gal. The blue colony color enhanced the ability to track the recombinant by dilution plating and lessened the reliance on antibiotic resistance as the sole selective agent. The objectives of the release were to evaluate the effectiveness of the *lac*ZY marking system and to compare the rhizosphere competence of the engineered strain and its parent. The bacteria were applied separately by spraying a suspension (5×10^8 to 1×10^9 cfu/ml) directly into the wheat seed furrow during planting. Three hours after planting, seeds were colonized by approximately 2×10^3 cfu/seed. Seven to 10 days after planting populations of both strains reached a maximum of 3×10^6 cfu/g and remained at that level for several weeks. By the fourth week after inoculation populations began a steady decline and by harvest, 31 weeks after planting, populations of PS3732RN and PS3732RNL11 were 2.3×10^2 and 4.6×10^2 cfu/g root, respectively. Lateral dissemination of the bacteria through the soil was negligible and limited to the first 18 cm from the point of application. Vertical dissemination was limited to a depth of 30 cm below the surface. These findings, along with the inability to detect transfer of the Tn*7*::*lac*ZY insert in any indigenous rhizosphere bacteria, indicated that minimal risk is associated with this type of release.

The *lac*ZY genes also were introduced into strain 2-79 to yield the recombinant 2-79RNL3. Both strains were applied to seeds of wheat (approximately 3 x 10^8 cfu/seed) and planted October 6, 1988 in a field near Pullman, Washington. The population trends of both strains were identical throughout the growing season. The largest populations, 10^8 cfu/g root, were recorded in the first 14 days after planting. Thereafter, populations of both strains declined and by the last sample, 13 days after harvest, the population had dropped by five orders of magnitude to about 10^3 cfu/g root. The population trends for 2-79 and 2-79RNL3 were very similar to those recorded for 2-79 in an earlier field study (39). No colonies of 2-79 or 2-79RNL3 were detected on roots of plants in nontreated rows 30 cm away from the inoculated rows. However, bacteria of both strains were recovered from the roots and seeds of volunteer (previous crop) lentil seedlings growing in the rows of inoculated wheat (77). This study also demonstrated that there was minimal risk of spread of the bacteria from the inoculated wheat.

1.8 Mechanisms of Biological Control by PGPR

In general, competition for nutrients supplied by roots and seeds and occupation of sites favored for colonization (niche exclusion) probably are responsible for a small to moderate degree of disease suppression by most PGPR and are of primary importance in some strains. Paulitz (78) reported that the biological control of *Pythium* damping-off by *P. putida* N1R, applied to soybean and pea seeds, was mediated through competition for seed volatiles which may serve as inducers and nutrients for *Pythium ultimum*. Similarly, *Enterobacter cloacae* may also annul the stimulatory activity of cotton seed volatiles to sporangia of *Pythium ultimum* (79,80). Of particular interest is the recent report by Wei et al. (81) that some PGPR applied to cucumber seeds induced a resistance response in the leaves to *Colletotrichum orbiculare*. Further, van Peer et al. (82) reported that *Pseudomonas* sp. WCS417r induced resistance in carnation to *Fusarium oxysporum* f.sp. *dianthi* and increased accumulation of phytoalexins in the stems. Like competition, induced resistance may also be a common underlying mechanism of biocontrol.

It has now been clearly demonstrated that for many PGPR, production of metabolites such as antibiotics, siderophores and hydrogen cyanide is the primary mechanism of biocontrol (7,8,9,68,83). Most recently, interest has focused on the secondary metabolites phenazine-1-carboxylic acid (PCA), 2,4-diacetylphloroglucinol (Phl), pyoluteorin (Plt) (84), pyrrolnitrin (85), oomycin A (86), and hydrogen cyanide (HCN). The basic strategy that is widely employed for determining the role of a specific gene or trait in a biocontrol process by PGPR involves: 1) development of an assay to demonstrate biocontrol activity; 2) selection of wild-type strains with biocontrol activity; 3) mutagenesis of strains; 4) screening of mutants for loss of the desired trait; 5) preparation of a genomic library from wild-type DNA; and 6) complementation of mutants to restore the desired trait (79). PCA and Phl currently are the most intensively studied metabolites.

Thomashow and Weller (87) provided the first conclusive evidence that production of antibiotics in situ contributes to biocontrol activity by PGPR. *Pseudomonas fluorescens* 2–79 produces PCA and suppresses take-all of wheat (88). Phenazine-deficient (Phz^-) Tn*5* mutants of 2–79 failed to inhibit *G. g. tritici* in vitro and were significantly less suppressive of take-all than 2–79. Mutants complemented with homologous DNA from a 2–79 genomic library were restored for production of PCA, inhibition of *G. g. tritici* and suppression of take-all. *P. chloroaphis* 30–84 also produces PCA, as well as, 2-hydroxyphenazine-1-carboxylic acid (2-OH-PCA) and 2-hydroxyphenazine (2-OH-PZ) and also suppresses take-all. Phz^- mutants, like those of 2–79, were noninhibitory to *G. g. tritici* and less suppressive of take-all than 30–84. Two overlapping cosmids from a genomic library of 30–84 DNA, each with identical *Eco*RI fragments of 11.2 kb and 9.2 kb, restored mutants to phenazine production and disease suppressiveness. *Escherichia coli* containing the 9.2 kb fragment produced all three phenazines. Two genes, *phz*B and *phz*C, involved in PCA and 2-OH-PCA production, respectively, were localized to a 2.8kb region of the 9.2 kb fragment (89). A putative activation gene *phz*A was identified upstream of *phz*B and *phz*C (90), (Pierson III personal communication), (91). Cosmid clones containing phenazine biosynthetic genes from 2–79 hybridized strongly with the 9.2 kb *Eco*RI fragment from 30–84. A 12-kb fragment containing the biosynthetic locus was sufficient to transfer PCA biosynthetic capability to other pseudomonads (91). Sequences required for PCA production in 2–79 were contained within two divergently transcribed units of approximately 5 kb and 0.75 kb that may correspond functionally to *phz*B and *phz*A, respectively, in 30–84.

As further support for the importance of phenazine antibiotics in control of take-all, PCA was isolated from the roots and rhizosphere of wheat treated with 2–79, 30–84 or their respective Phz^+ complemented mutants and grown in raw soil (28–133 ng PCA/g of root with adhering soil). PCA was not recovered from roots of nontreated wheat or wheat treated with Phz^- Tn*5* mutants (92).

P. fluorescens CHA0 produces at least five bioactive compounds including Phl, Plt, HCN, indoleacetic acid and a fluorescent siderophore. Extensive studies have been conducted to identify the role of each in the suppression of black root rot of tobacco, caused by *Thielaviopsis basicola*, take-all and damping-off of cucumber caused by *Pythium ultimum*. Using the genetic approach described above it has been demonstrated that production of Phl is the primary mechanism of take-all suppression and that both Phl and HCN contribute to biocontrol of black root rot (93,94,95,96). Plt apparently contributes to the suppression of damping-off (97). Keel et al. (94) isolated Phl from wheat roots colonized by CHA0 (0.94–1.36 μg Phl/g root).

Phl also has been shown to contribute to the suppression of take-all by *P. aureofaciens* Q2–87 (98,99) and *Pythium* damping off of sugar beet by *Pseudomonas* strain F113 (100,101). From both of these strains putative Phl biosynthetic loci have been cloned.

1.9 Inconsistant Performance of PGPR

Inconsistant performance in the field is the major impediment to the large-scale commercial development and use of PGPR in agriculture. Performance of an introduced strain often varies from site to site and year to year. Many factors can contribute to the inconsistent performance of PGPR given the complex interactions among the host, pathogen, bacteria and the soil environment (9). One of the most important factors is variability in root colonization. Given the multiple steps involved in the process it should not be surprising that colonization often is erratic. Introduced bacteria become lognormally rather than normally distributed among roots (16) and root systems (24) meaning that population sizes from root to root can vary by several orders of magnitude and some roots may be completely unprotected. The demonstration by Bull et al. (22) of an inverse relationship between the population size of 2–79 on wheat roots and the number of take-all lesions on the same roots, underscores the fact that incomplete colonization will reduce the chances for successful biocontrol.

Another important factor is inconsistent production or inactivation in situ of the secondary metabolites that contribute to disease control. For control to occur, production of these metabolites must coincide with the period of time when the plant is vulnerable to attack. However, production of secondary metabolites, such as phenazines, is highly dependant on cultural conditions (102,103). In the rhizosphere, the temporal regulation of metabolically expensive secondary metabolites is likely to be even more tightly controlled and very dependent on the environment within the microsite. For example, oomycin A biosynthesis was induced by glucose but inhibited by combinations of amino acids, all of which are found in root exudates. Further, both temperature and water potential affects expression of *afu*E, a key oomycin A biosynthetic gene, in the rhizosphere of cotton (26).

Ownley et al. (104) found that the performance of 2–79 against take-all varied considerably in 10 soils. It was shown that of 28 soil variables determined for these 10 soils, seven including ammonium-nitrogen, sulfate-sulfur, zinc, soil pH, extractable and soluble sodium and percent sand, were directly related to the biocontrol activity of 2–79. In contrast, nine variables including cation exchange capacity, percent silt, percent clay, exchangeable acidity, manganese, iron, percent organic matter, total carbon and total nitrogen were inversely related to biocontrol activity. The positive correlation of some variables such as ammonium-nitrogen, sulfate-sulfur and zinc to enhanced biocontrol activity is speculated to result from an effect on phenazine production in the rhizosphere. On the other hand, variables such as clay, silt, organic matter may be involved in the tie-up of PCA after it is produced. The importance of zinc to in situ PCA production was further suggested by the finding that take-all suppression by 2–79 was significantly greater in a Woodburn silt loam (naturally low in zinc) amended with 50 ug zinc (as Zn-EDTA/g soil) than in the nonamended soil (B.H. Ownley and D.M. Weller, unpublished). Interestingly, in a study of the nutritional requirements of both cell growth and PCA production by 2–79 in liquid culture, $ZnSO_4$ enhanced antibiotic production without increasing cell growth (105).

1.10 Improving Root Colonizing and Biological Control

One approach to increasing root colonization by PGPR is to increase the dose of the bacteria applied to the seed. Bull et al. (22) showed that the size of the population of strain 2–79 that became established on wheat roots grown in raw soil was directly related to the initial population applied to the seed. Similar dose effects were reported for *Azospirillum* applied to wheat (106) and *P. fluorescens* applied to potato (16). However, increasing colonization by increasing the initial dose of bacteria on the seed has limitations (107). In wheat, for example, populations of certain Phl or Phz producing pseudomonads can be phytotoxic at concentrations above 5 x 10^8 cfu/seed. Furthermore, although a larger dose will increase root populations, the frequency of roots colonized may not be increased. Finally, increasing the dose may substantially increase the cost of a treatment (108).

Another approach to increase colonization and biocontrol is the application of mixtures of strains. PGPR research has focused primarily on the use of single strains. However, Weller and Cook (109) demonstrated that *P. fluorescens* 2–79 used in combination with *P. fluorescens* 13–79 was superior to either strain alone in about 50% of the trials. More recently, mixtures of fluorescent *Pseudomonas* strains Q2–87 + Qlc-80 + Q8d-80 + Q65c-8 and Q2–87 + Qlc-80 + Q8d-80 + Q69c-80 have provided significantly more control of take-all and greater increase in yield in field studies than each strain used individually. For example, in a spring wheat trial the combination Q2–87 + Qlc-80 + Q8d-80 + Q65c-80 increased yield 20% over the nontreated control, but individual strains increased yield no more than 5% (110). It is hypothesized that the greater diversity of phenotypes associated with combinations results in a diverse and potentially more stable rhizosphere community that is able to more thoroughly colonize roots and/or survive the biological, chemical and physical changes that occur in the rhizosphere throughout the growing season. Secondly, mixtures may provide a more diverse "arsenal" of mechanisms capable of suppressing both target and nontarget pathogens. Thus, with multiple strains there is a greater probability that at least some of the genes involved in biocontrol will be expressed over a wider range of environmental condition and microhabitats.

Recombinant DNA technology has provided the most exciting and potentially successful means to improve root colonization and biological control by PGPR. One approach involves enhancing traits that are important. For example, Maurhofer et al. (97) demonstrated that introduction of the cosmid pME3090 (carrying a 22 kb insert of CHA0 DNA) into CHA0 enhanced the production of Plt. The recombinant protected cucumber against damping-off caused by *Pythium ultimum* better than the parental strain. Production of oomycin A and suppression of *Pythium* damping-off of cotton were increased by placing the oomycin A biosynthetic gene cluster in *P. fluorescens* Hv37a under the control of the constitutive *tac* promotor from *E. coli.*

Improvements can also be achieved through the transfer of biocontrol traits into other strains. Kerr (111) earlier warned about over-optimism with this approach, however, it is now clear that transfer and heterologous expression of important biocontrol genes (particularly those involved in antibiotic production) are more easily achieved than previously anticipated. For example, transfer of pCU203 containing a putative Phl biosynthetic locus from *Pseudomonas* strain F113 into M114 resulted in a recombinant that produced Phl and suppressed *Pythium* damping-off of sugar beet better than M114 (100). Similarly, transfer of Phl biosynthetic genes from Q2-87 into *Pseudomonas* strains 2-79 and 5097 resulted in the ability to synthesize Phl and increased their inhibition of *G. g. tritici*, *P. ultimum* and *Rhizoctonia solani* (98). Introduction of pPH2108A, containing the phenazine biosynthetic locus from 2-79, resulted in PCA production by all 27 fluorescent *Pseudomonas* spp. into which it was mobilized, some of the recombinants showed enhanced suppression of take-all (H. Hara, L.S. Thomashow, D.M. Weller and D.E. Essar, unpublished data).

1.11 Conclusion

There has been considerable research and progress in the last 15 years in the understanding of the process of root colonization and mechanisms of growth-promotion by PGPR. Although, PGPR technology is slowly being introduced into agriculture, many impediments still exist to the widespread commercialization and use of PGPR. PGPR research must intensify because this technology will be challenged in the near future to fill a void as the use of chemical pesticides become more restricted in agriculture. Emphasis is needed on identifying soil edaphic factors that affect biocontrol activity and root colonization, as well as, those traits that contribute to rhizosphere competence. Also critical is the development of uniform and scientifically based guidelines for the release of genetically engineered PGPR in order to facilitate more routine screening in the field.

1.12 References

1. Kloepper, J. W., B. Schippers, and P.A.H.M. Bakker. Proposed elimination of the term *endorhizosphere*. Phytopathology. *82* (1992) 726-727.
2. Curl, E.A., and B. Truelove. The Rhizosphere, Springer-Verlag, Berlin 1986.
3. Foster, R.C., A.D. Rovira, and T.W. Cock. Ultrastructure of the root-soil interface. Am. Phytopathol. Soc., St. Paul. 1983.
4. Kloepper J.W., R. Lifshitz, M.N. Schroth. *Pseudomonas* inoculants to benefit plant production. In: ISI Atlas of Science. Institute for Scientific Information, Philadelphia 1988, pp. 60-64.
5. Schroth, M.N. and J.G. Hancock. Selected topics in biological control. Annu. Rev. Microbiol. *35* (1981) 453-476.
6. Schroth, M.N. and J.G. Hancock. Disease-suppressive soil and root colonizing bacteria. Science *216* (1982) 1376-1381.
7. Lugtenberg, B.J.J., L.A. de Weger, and J.W. Bennett. Microbial stimulation of plant growth and protection from disease. Current Opinion in Microbiology *2* (1991) 457-464.
8. O'Sullivan, D.J., and F. O'Gara. Traits of fluorescent *Pseudomonas* spp. involved in suppression of plant root pathogens. Microbiol. Rev. *56* (1992) 662-676.

9. Weller, D.M. Biological control of soilborne pathogens in the rhizosphere with bacteria. Ann. Rev. Phytopathol. *26* (1988) 379-407.
10. Salt, G.A. The increasing interest in "minor pathogen". In: B. Schippers and W. Gams (Eds.), Soil-Borne Plant Pathogens. Academic, London/New York/SanFrancisco 1979, pp. 209-227.
11. Schippers, B., A.W. Bakker, and P.A.H.M. Bakker. Interactions of deleterious and beneficial rhizosphere microorganisms and the effect of cropping practice. Ann. Rev. Phytopathol. *25* (1987) 339-358.
12. Suslow, T.V., and M.N. Schroth. Role of deleterious rhizobacteria as minor pathogens in reducing crop growth. Phytopathology. *72* (1982) 111-115.
13. Parke, J.L. Root colonization by indigenous and introduced microorganisms. In: D.L. Keister and P.B. Cregan (Eds.), The rhizosphere and plant growth. Kluwer Academic Publishers, Dordrecht 1991, pp. 33-42.
14. Deacon, J.W. Significance of ecology in the development of biocontrol agents against soil-borne plant pathogens. Biocontrol Science and Technology *1* (1991) 5-20.
15. Liddell, C.M., and J.L. Parke. Enhanced colonization of pea tap roots by a fluorescent pseudomonad biocontrol agent by water infiltration into soil. Phytopathology *79* (1989) 1327-1332.
16. Loper, J.E., C. Haack, M.N. Schroth. Population dynamics of soil pseudomonads in the rhizosphere of potato (*Solanum tuberosum L.*). Appl. Environ. Microbiol. *49* (1985) 416-422.
17. Suslow, T.V. Role of root-colonizing bacteria in plant growth. In: M.S. Mount, G.H. Lacy (Eds.), Phytopathogenic Prokaryotes. Academic, London 1982, pp. 187-223.
18. Caesar, A.J., and T.J. Burr. Growth promotion of apple seedlings and root-stocks by specific strains of bacteria. Phytopathology *77* (1987) 1583-1588.
19. Kloepper, J.W., and M.N. Scroth. Relationship of *in vitro* antibiosis of plant growth-promoting rhizobacteria to plant growth and the displacement of root microflora. Phytopathology *71* (1981) 1020-1024.
20. Yuen G.Y., M.N. Schroth. Interactions of *Pseudomonas fluorescens* strain E6 with ornamental plants and its effect on the composition of root-colonizing microflora. Phytopathology *76* (1986) 176-180.
21. Xu, G.-W., and D.C. Gross. Field evaluations of the interactions among fluorescent pseudomonads, *Erwinia carotovora*, and potato yields. Phytopathology *76* (1986) 423-430.
22. Bull, C.T., D.M. Weller, and L.S. Thomashow. Relationship between root colonization and suppression of *Gaeumannomyces graminis* var.*tritici* by *Pseudomonas fluorescens* strain 2-79. Phytopathology *81* (1991) 954-959.
23. Howie, W.J., R.J. Cook, and D.M. Weller. Effects of soil matric potential and cell motility on wheat root colonization by fluorescent pseudomonads suppressive to take-all. Phytopathology *77* (1987) 286-292.
24. Bahme, J.B., and M.N. Schroth. Spatial-temporal colonization patterns of a rhizobacterium on underground organs of potato. Phytopathology *77* (1987) 1093-1100.
25. Bull, C.T. Wheat root colonization by disease-suppressive or nonsuppressive bacteria and the effect of population size on severity of take-all caused by *Gaeumannomyces graminis* var. *tritici*. M.S. Thesis, Wash. State University, Pullman, 1987, pp. 75.
26. Gutterson N., W. Howie, and T. Suslow. Enhancing effects of biocontrol agents by use of biotechnology. In: R.R. Baker and P.E. Dunn (Eds.), New Directions in biological control: Alternatives for suppressing agricultural pests and diseases. Alan R. Liss, Inc., New York 1990, pp. 749-765.
27. Seong, K-Y, M. Hofte, J. Boelens, and W. Verstraete. Growth, survival, and root colonization of plant growth beneficial *Pseudomonas fluorescens* ANP15 and *Pseudomonas aeruginosa* 7NSK2 at different temperatures. Soil Biol. Biochem. *23* (1991) 423-428.
28. Vandenhove, H., R. Merckx, H. Wilmots, and K. Vlassak. Survival of *Pseudomonas fluorescens* inocula of different physiological stages in soil. Soil Biol. Biochem. *23* (1991) 1133-1142.
29. Howie, W.J. Factors affecting colonization of wheat roots and suppression of take-all by pseudomonads antagonistic to *Gaeumannomyces graminis* var. *tritici*. Ph.D. Dissertation. Wash. State. Univ., Pullman, 1985.
30. Bowen, G.D., and A.D. Rovira. Microbial colonization of plant roots. Ann. Rev. Phytopathol. *14* (1976) 121-144.
31. Stephens, P.M., M. O'Sullivan, and F. O'Gara. Effect of bacteriophage on colonization of sugarbeet roots by fluorescent *Pseudomonas* spp. Appl. Environ. Microbiol. *53* (1987) 1164-1167.
32. Casida, L.E. Competitive ability and survival in soil of *Pseudomonas* strain 679-2, a dominant, nonobligate bacterial predator of bacteria. Appl. Environ. Microbiol. *58* (1992) 32-37.

33. Shikano, S., L.S. Luckinbill, and Y. Kurihara. Changes of traits in a bacterial population associated with protozoal predation. Microb. Ecol. *20* (1990) 75–84.
34. Kremer, R.J., M.F.T. Begonia, L. Stanley and E.T. Lanham. Characterization of rhizobacteria associated with weed seedlings. Appl. Environ. Microbiol. *56* (1990) 1649–1655.
35. Atkinson, T.G., J.L. Neal Jr. and R.I. Larson. Genetic control of the rhizosphere microflora of wheat. In: G.W. Bruehl (Ed.), Biology and control of soil-borne plant pathogens. Am. Phytopathol. Soc., St. Paul 1975, pp. 116–122.
36. Azad, H.R., J.R. Davis, W.C. Schnathorst and C.I. Kado. Relationship between rhizoplane and rhizosphere bacteria and verticillium wilt resistance in potato. Arch. Microbiol. *140* (1985) 347–351.
37. Miller, H.J., G. Henken, J.A. van Veen. Variation and composition of bacterial populations in the rhizospheres of maize, wheat, and grass cultivars. Can. J. Microbiol. *35* (1989) 656–660.
38. Weller, D.M. Effects of wheat genotype on root colonization by a take-all suppressive strain of *Pseudomonas fluorescens*. Phytopathology. *76* (1986) 1059 (Abstr.).
39. Weller, D.M. Colonization of wheat roots by a fluorescent pseudomonad suppressive to take-all. Phytopathology. *73* (1983) 1548–1553.
40. Mazzola, M. and R.J. Cook. Effects of fungal root pathogens on the population dynamics of biocontrol strains of fluorescent pseudomonads in the wheat rhizosphere. Appl. Environ. Microbiol. *57* (1991) pp. 2171–2178.
41. James, D.W. Jr., T.V. Suslow, and K.E. Steinback. Relationship between rapid, firm adhesion and long-term colonization of roots by bacteria. Appl. Environ. Microbiol. *50* (1985) 392–397.
42. Anderson, A.J., P. Habibzadegah-Tari, and C.S. Tepper. Molecular studies on the role of a root surface agglutinin in adherence and colonization by *Pseudomonas putida*. Appl. Environ. Microbiol. *54* (1988) 375–380.
43. Tari, P.H., A.J. Anderson. Fusarium wilt suppression and agglutinability of *Pseudomonas putida*. Appl. Environ. Microbiol. *54* (1988) 2037–2041.
44. Buell, C.R., and A.J. Anderson. Genetic analysis of the *agg*A locus involved in agglutination and adherence of *Pseudomonas putida*, a beneficial fluorescent pseudomonad. Mol. Plant-Microbe Interact. *5* (1992) 154–162.
45. van Peer, R., H.L.M. Punte, L.A. de Weger, and B. Schippers. Characterization of root surface and endorhizosphere pseudomonads in relation to their colonization of roots. Appl. Environ. Microbiol. *56* (1990) 2462–2470.
46. Chao, W-L., R-K. Li, W-T. Chang. Effect of root agglutinin of microbial activities in the rhizosphere. Appl. Environ. Microbiol. *54* (1988) 1838–1841.
47. De Mot, R., H. Joos, A. Van Gool, and J. Vanderleyden. Colonization of wheat roots by *Pseudomonas fluorescens*: Scanning electron microscopy and biochemical analysis. In: D.L. Keister and P.B. Cregan (Eds.), The rhizosphere and plant growth. Kluwer Academic Publishers, Dordrect, 1991.
48. De Mot, R., and J. Vanderleyden. Purification of a root-adhesive outer membrane protein of root-colonizing *Pseudomonas fluorescens*. FEMS Microbiology Letters. *81* (1991) 323–328.
49. De Mot, R., P. Proost, J. van Damme, and J. Vanderleyden. Homology of the root adhesin of *Pseudomonas fluorescens* OE 28.3 with porin F of *P. aeroginosa* and *P. syringae*. Mol. Gen. Genet. *231* (1992) 489–493.
50. Finnen, R.L., N.L. Martin, R.J. Siehnel, W.A. Woodruff, M. Rosok and R.E.W. Hancock. Analysis of the *Pseudomonas aeruginosa* major outer membrane protein OprF by use of truncated OprF derivatives and monoclonal antibodies. J. Bacteriol. *174* (1992) 4977–4985.
51. Douglas, C.J., W. Halperin, and E.W. Nester. *Agrobacterium tumefaciens* mutants affected in attachment to plant cells. J. Bacteriol. *152* (1982) 1265–1275.
52. Douglas, C.J., R.J. Staneloni, R.A. Rubin, E.W. Nester. Identification and genetic analysis of an *Agrobacterium tumefaciens* chromosomal virulence region. J. Bacteriol. *161* (1985) 850–860.
53. Thomashow, M.F., J.E. Karlinsey, J.R. Marks, R.E. Hurlbert. Identification of a new virulence locus in *Agrobacterium tumefaciens* that affects polysaccharide composition and plant cell attachment. J. Bacteriol. *169* (1987) 3209–3216.
54. Cangelosi, G.A., L. Hung, V. Puvanesarajah, G. Stacey, D.A. Ozga, J.A. Leigh, and E.W. Nester. Common loci for *Agrobacterium tumefaciens* and *Rhizobium meliloti* exopolysaccharide synthesis and their roles in plant interactions. J. Bacteriol. *169* (1987) 2086–2091.
55. Leigh, J.A., E.R. Signer and G.C. Walker. Exopolysaccharide-deficient mutants of *Rhizobium meliloti* that form ineffective nodules. Proc. Natl. Acad. Sci. USA *82* (1985) 6231–6235.

56. Waelkens, F., M. Maris, C. Verreth, J. Vanderleyden and A. Van Gool. *Azospirillum* DNA shows homology with *Agrobacterium* chromosomal virulence genes. FEMS Microbiol. Lett. *43* (1987) 241-246.
57. Dobereiner, J. and F.O. Pedrosa. Nitrogen-fixing bacteria in non-leguminous crop plants. Science Tech., Madison, 1987.
58. Michiels, K.W., J. Vanderleyden, A.P. Van Gool, and E.R. Signer. Isolation and characterization of *Azospirillum brasilense* loci that correct *Rhizobium meliloti exo*B and *exo*C mutations. J. Bacteriol. *170* (1988) 5401-5404.
59. Haahtela, K., and T.K. Korhonen. In vitro adhesion of N_2-fixing enteric bacteria to roots of grasses and cereals. Appl. Environ. Microbiol. *49* (1985) 1186-1190.
60. Haahtela, K., E. Tarkka and T.K. Korhonen. Type 1 fimbria-mediated adhesion of enteric bacteria to grass roots. Appl. Environ. Microbiol. *49* (1985) 1182-1185.
61. Korhonen, T.K., E.-L. Nurmiaho-Lassila, T. Laakso and K. Haahtella. Adhesion of fimbriated nitrogen-fixing enteric bacteria to roots of grasses and cereals. Plant and Soil. *90* (1986) 59-69.
62. Korhonen, T.K., E. Tarkka, H. Ranta, and K. Haahtella. Type 3 fimbriae of *Klebsiella* sp.: molecular characterization and role in bacterial adhesion to plant roots. J. Bacteriol. *155* (1983) 860-865.
63. Vesper, S.J. Production of pili (fimbriae) by *Pseudomonas fluorescens* and correlation with attachment to corn roots. Appl. Environ. Microbiol. *53* (1987) 1397-1403.
64. Vesper, S.J. and W.D. Bauer. Role of pili (fimbriae) in attachment of *Bradyrhizobium japonicum* to soybean roots. Appl. Environ. Microbiol. *52* (1986) 134-141.
65. Vesper, S.J., N.S.A. Malik and W.D. Bauer. Transposon mutants of *Bradyrhizobium japonicum* altered in attachment to host roots. Appl. Environ. Microbiol. *53* (1987) 1959-1961.
66. Scher, F.M., J.W. Kloepper, C. Singleton, I. Zaleska, and M. Laliberte. Colonization of soybean roots by *Pseudomonas* and *Serratia* species: relationship to bacterial motility, chemotaxis, and generation time. Phytopathology. *78* (1988) 1055-1059.
67. De Weger, L.A., C.I.M. van der Vlugt, A.H.M. Wijfjes, P.A.H.M. Bakker, B. Schippers, B. Lugtenberg. Flagella of a plant-growth-stimulating *Pseudomonas fluorescens* strain are required for colonization of potato roots. J. Bacteriol. *169* (1987) 2769-2773.
68. Weller, D.M., and L.S. Thomashow. Advances in rhizobacteria for biocontrol. Current Opinion in Biotechnology. *4* (1993) 306-311.
69. Mazzola, M., R.J. Cook, L.S. Thomashow, D.M. Weller, and L.S. Pierson III. Contribution of phenazine antibiotic biosynthesis to the ecological competence of fluorescent pseudomonads in soil habitats. Appl. Environ. Microbiol. *58* (1992) 2616-2624.
70. Scher, F.M., J.W. Kloepper and C.A. Singleton. Chemotaxis of fluorescent *Pseudomonas* spp. to soybean seed exudates *in vitro* and in soil. Can. J. Microbiol. *31* (1985) 570-574.
71. Heinrich, D., and D. Hess. Chemotactic attraction of *Azospirillum lipoferum* by wheat roots and characterization of some attractants. Can. J. Microbiol. *31* (1985) 26-31.
72. Ahmad, J.S., and R. Baker. Rhizosphere competence of *Trichoderma harzianum*. Phytopathology. *77* (1987) 182-189.
73. Ahmad, J.S., and R. Baker. Competitive saprophytic ability and cellulotytic activity of rhizosphere-competent mutants of *Trichoderma harzianum*. Phytopathology. *77* (1987) 358-362.
74. Weller, D.M. Distribution of a take-all suppressive strain of *Pseudomonas fluorescens* on seminal roots of winter wheat. Appl. Environ. Microbiol. *48* (1984) 897-899.
75. Kluepfel, D.A., E.L. Kline, H.D. Skipper, T.A. Hughes, D.T. Gooden, D.J. Drahos, G.F. Barry, B.C. Hemming, and E.J. Brandt. The release and tracking of genetically engineered bacteria in the environment. Pythopathology. *81* (1991) 348-352.
76. Drahos, D.J., B.C. Hemming, and S. McPherson. Tracking recombinant organisms in the environment: β-galactosidase as a selectable non-antibiotic marker for fluorescent pseudomonads. Bio/Technology *4* (1986) 439-444.
77. Cook, R.J., D.M. Weller, P. Kovacevich, D. Drahos, B. Hemming, G. Barnes, and E.A. Pierson. Establishment, monitoring, and termination of field tests with genetically altered bacteria applied to wheat for biological control of take-all. In: The biosafety results of field tests of genetically modified plants and microorganisms. Agricultural Research Institute, Bethesda 1990, pp. 177-187.
78. Paulitz, T.C. Effect of *Pseudomonas putida* on the stimulation of *Pythium ultimum* by seed volatiles of pea and soybean. Phytopathology. *81* (1991) 1282-1287.

79. Nelson, E.B., and A.P. Maloney. Molecular approaches for understanding biological control mechanisms in bacteria: studies of the interaction of *Enterobacter cloacae* with *Pythium ultimum*. Can. J. Plant Pathology. *14* (1992) 106-114.
80. Nelson, E.B. Exudate molecules initiating fungal responses to seeds and roots. Plant and Soil. *129* (1990) 61-73.
81. Wei, G., J.W. Kloepper, and S. Tuzun. Induction of systematic resistance of cucumber to *Colletotrichum orbiculare* by select strains of plant growth-promoting rhizobacteria. Phytopathology. *81* (1991) 1508-1512.
82. van Peer, R., G.J. Niemann, and B. Schippers. Induced resistance and phytoalexin accumulation in biological control of fusarium wilt of carnation by *Pseudomonas* sp. strain WCS417r. Phytopathology. *81* (1991) 728-734.
83. Leong, J. Siderophores: their biochemistry and possible role in the biocontrol of plant pathogens. Annu. Rev. Phytopathol. *24* (1986) 187-209.
84. Howell, C.R., and R.D. Stipanovic. Suppression of *Pythium ultimum*-induced damping-off of cotton seedlings by *Pseudomonas fluorescens* and its antibiotic, pyoluteorin. Phytopathology. *70* (1980) 712-715.
85. Howell, C.R., and R.D. Stipanovic. Control of *Rhizoctonia solani* on cotton seedlings with *Pseudomonas fluorescens* and with an antibiotic produced by the bacterium. Phytopathology. *69* (1979) 480-482.
86. Howie, W.J., and T.V. Suslow. Role of antibiotic biosynthesis in the inhibition of *Pythium ultimum* in the cotton spermosphere and rhizosphere by *Pseudomonas fluorescens*. Mol. Plant-Microbe Interact. *4* (1991) 393-399.
87. Thomashow, L.S., and D.M. Weller, Role of a phenazine antibiotic from *Pseudomonas fluorescens* in biological control of *Gaeumannomyces graminis* var. *tritici*. J. Bacteriol. *170* (1988) 3499-3508.
88. Cook, R.J., and D.M. Weller. Management of take-all in consecutive crops of wheat or barley. In: I. Chet (Ed.), Innovative approaches to plant disease control. John Wiley & Sons, Inc., New York 1986, pp. 41-76.
89. Pierson III, L.S., and L.S. Thomashow. Cloning and heterologous expression of the phenazine biosynthetic locus from *Pseudomonas aureofaciens* 30-84. Mol. Plant-Microbe Interact. *5* (1992) 330-339.
90. Pierson III, L.S., and V.D. Keppenne. Identification of a locus that acts *in trans* to stimulate phenazine gene expression in *Pseudomonas aureofaciens* 30-84. Abstract 197, 6th International Symposium on Molecular Plant-Microbe Interactions, Seattle, 1992.
91. Thomashow, L.S., D.W. Essar, D.K. Fujimoto, L.S. Pierson III, C. Thrane, and D.M. Weller. Genetic and biochemical determinants of phenazine antibiotic production in fluorescent pseudomonads that suppress take-all disease of wheat. In: E.W. Nester and D.P.S. Verma (Eds.), Advances in molecular genetics of plant-microbe interactions, Vol 2. Kluwer Academic Publishers, Dordrecht 1993, 535-541.
92. Thomashow, L.S., D.M. Weller, R.F. Bonsall and L.S. Pierson III. Production of the antibiotic Phenazie-1-Carboxylic acid by fluorescent *Pseudomonas* species in the rhizophere of wheat. Appl. environ. microbiol. *56* (1990) 908-912.
93. Haas, D., C. Keel, J. Laville, M. Maurhofer, T. Oberhansli, U. Schnider, C. Voisard, B. Wüthrich, and G. Defago. Secondary metabolites of *Pseudomonas fluorescens* strain CHA0 involved in the suppression of root diseases. In: H. Hennecke and D.P.S. Verma (Eds.), Advances in molecular genetics of plant-microbe interactions. Kluwer Academic Publishers, Dordrecht 1991, pp. 450-456.
94. Keel, C., U. Schnider, M. Maurhofer, C. Voisard, J. Laville, U. Burger, P. Wirthner, D. Haas, and G. Defago. Suppression of root diseases by *Pseudomonas fluorescens* CHA0: importance of the bacterial secondary metabolite 2,4-diacetylphloroglucinol. Mol. Plant-Microbe Interact. *5* (1992) 4-13.
95. Keel, C., P. Wirthner, T. Oberhansli, C. Voisard, U. Burger, D. Haas, and G. Defago. Pseudomonads as antagonists of plant pathogens in the rhizosphere: role of the antibiotic 2,4-diacetylphloroglucinol in the suppression of black root rot of tobacco. Symbiosis. *9* (1990) 327-341.
96. Voisard, C., C. Keel, D. Haas, and G. Defago. Cyanide production by *Pseudomonas fluorescens* helps suppress black root rot of tobacco under gnotobiotic conditions. EMBO J. *8* (1989) 361-358.
97. Maurhofer, M., C. Keel, U. Schnider, C. Voisard, D. Haas, and G. Defago. Influence of enhanced antibiotic production in *Pseudomonas fluorescens* strain CHA0 on its disease suppressive capacity. Phytopathology. *82* (1992) 190-195.

98. Vincent, M.N., L.A. Harrison, J.M. Brackin, P.A. Kovacevich, P. Mukerji, D.M. Weller, and E.A. Pierson. Genetic analysis of the antifungal activity of a soilborne *Pseudomonas aureofaciens* strain. Appl. Environ. Microbiol. *57* (1991) 2928–2934.
99. Harrison, L.A., L. Letendre, P. Kovacevich, E. Pierson, and D. Weller. Purification of an antibiotic effective against *Gaeumannomyces graminis* var. *tritici* produced by a biocontrol agent, *Pseudomonas aureofaciens*. Soil Biol. Biochem. *25* (1993) 215–221.
100. Fenton, A.M., P.M. Stephens, J. Crowley, M. O'Callaghan, and F. O'Gara. Exploitation of gene(s) involved in 2,4-diacetylphloroglucinol biosynthesis to confer a new biocontrol capability to a *Pseudomonas* strain. Appl. Environ. Microbiol. *58* (1992) 3873–3878.
101. Shanahan, P., D.J. O'Sullivan, P. Simpson, J.D. Glennon, and F. O'Gara. Isolation of 2,4-diacetylphloroglucinol from a fluorescent pseudomonad and investigation of physiological parameters influencing its production. Appl. Environ. Microbiol. *58* (1992) 352–358.
102. Kanner, D., N.N. Gerber, and R. Bartha. Pattern of phenazine pigment production by a strain of *Pseudomonas aeruginosa*. J. Bacteriol. *134* (1978) 690–692.
103. Thomashow, L.S., and Pierson III, L.S. Genetic aspects of phenazine antibiotic production by fluorescent pseudomonads that suppress take-all disease of wheat. *In* H. Hennecke and D.P.S. Verma (Eds.), Advances in molecular genetics of plant-microbe interactions. Kluwer Academic Publishers, Dordrecht 1991, 443–449.
104. Ownley, B.H., D.M. Weller, and J.R. Alldredge. Relation of soil chemical and physical factors with suppression of take-all by *Pseudomonas fluorescens* 2–79. In: C. Keel, B. Koller and G. Defago (Eds.), Plant growth-promoting rhizobacteria-progress and prospects. WPRS Bulletin, 1991/XIV/8, pp. 299–301.
105. Slininger, P.J., and M.A. Jackson. Nutritional factors regulating growth and accumulation of phenazine 1-carboxylic acid by *Pseudomonas fluorescens* 2–79. Appl. Microbiol. Biotechnol. *37* (1992) 388–392.
106. Bashan, Y. Migration of the rhizosphere bacteria *Azospirillium brasilense* and *Pseudomonas fluorescens* toward wheat roots in soil. J. Gen. Microbiol. *132* (1986) 3407–3414.
107. Osburn, R.M., M.N. Schroth, J.G. Hancock, and M. Hendson. Dynamics of sugar beet seed colonization by *Pythium ultimum* and *Pseudomonas* species: effects on seed rot and damping-off. Phytopathology. *79* (1989) 709–716.
108. Baker, C.A. and J.M.S. Henis. Commercial production and formulation of microbial biocontrol agents. In: R.R. Baker and P.E. Dunn (Eds.), New Directions in Biological Control: Alternatives for Suppressing Agricultural Pests and Diseases. Alan R. Liss, New York 1990, pp. 333–334.
109. Weller, D.M., and R.J. Cook. Suppression of take-all of wheat by seed treatments with fluorescent pseudomonads. Phytopathology. *73* (1983) 463–469.
110. Pierson, E.A., and D.M. Weller. Recent work on control of take-all of wheat by fluorescent pseudomonads. In: C. Keel, B. Koller and G. Defago (Eds.), Plant growth-promoting rhizobacteria-progress and prospects, WPRS Bulletin 1991/XIV/8, pp. 96–97.
111. Kerr, A. The impact of molecular genetics on plant pathology. Ann. Rev. Phytopathol. *25* (1987) 87–110.

2 Studies on Indigenous Endophytic Bacteria of Sweet Corn and Cotton

John A. McInroy and Joseph W. Kloepper

2.1 Introduction

Several reports since 1948 have demonstrated that bacteria naturally inhabit healthy plant tissues, including fruits (1), vegetables (2, 3), stems (4) and roots (5). Mundt and Hinkle (6) found endophytic bacteria within seeds and ovules of 25 of 27 plant species sampled, establishing the presence of endophytes prior to germination. Endophytic bacteria were found throughout cotton plants, radicles, roots, stems, unopened flowers, and bolls (7). Since bacterial endophytes have a natural association with plants and can colonize plant tissues without inciting disease, they are potenial candidates for use as agricultural inoculants which provide plant growth-promotion or biological control of plant diseases.

To date, there has been little research aimed at determining possible benefits of endophytic bacteria on crops. Van Peer et al. (8) reported that 30% of *Pseudomonas* endophytes reduced plant growth after seed bacterization and 2% actually stimulated plant growth. Researchers at Crop Genetics International have modified a xylem-inhabiting endophyte of bermudagrass (*Cynodon dactylon*), *Clavibacter xyli* subsp. *cynodontis*, using recombinant DNA techniques to produce an endotoxin from *Bacillus thuringiensis*, which combats the European corn borer (*Ostrinia nubilalis*) in corn (9). These two reports demonstrate that select endophytic bacterial strains may benefit plants.

It is possible that some endophytes can systematically colonize plants and overcome the limitations of phylloplane or rhizosphere bacteria. The internal tissues of plants provide a more uniform and protective environment for introduced biological control agents compared to the phylloplane, where exposure to ultraviolet radiation, rainfall and temperature fluctuations negatively affects introduced microorganisms, or compared to the rhizosphere where introduced microorganisms compete for nutrients with other microbes. Reports on the extent and density of tissue colonization by endophytic bacteria are limited, especially for above-ground tissues. It is therefore useful to have a quantitative understanding of the indigenous endophytic bacterial community to help assess endophytes as potential sources of effective strains for plant growth-promotion or biological control of plant disease.

The objectives of this study were to determine the population dynamics of bacterial endophytes in stems and roots during the growing season; to identify the major taxa of the endophytic community; and to compare populations in a model monocotyledonous plant, sweet corn (*Zea mays L.*) and a dicotyledonous plant, cotton (*Gossypium hirsutum L.*).

2.2 Materials and Methods

2.2.1 Media

Bacteria were isolated on three different media; tryptic soy agar (TSA) (Difco Laboratories, Detroit, MI) was used to support the growth of a broad range of microorganisms; medium R2A (Difco Laboratories, Detroit, MI) was used for bacteria requiring a low level of nutrients (oligotrophs); medium SC (10) was included to support the growth of some fastidious organisms, e.g. *Clavibacter xyli.*

2.2.2 Field Experiments

Cotton ("DES 119") and sweet corn ("Silver Queen") were planted in 1990 in the field in a fine-loamy, siliceous, thermic, Typic Hapludult soil at Tallassee, AL. Ten blocks of four 25-ft rows were planted for both crops. Plants were sampled at emergence and 2, 7, 14, 21, 28, 42, 56, 70 and 112 days after emergence. The experiment was repeated the following year under the same conditions and sampled once prior to emergence, at emergence and 7, 14, 28, 42, 56 and 70 days after emergence. At each sampling date, one randomly selected cotton plant from each of the ten replicate blocks was manually uprooted and transported at 10 °C to the laboratory. Sweet corn was sampled similarly.

2.2.3 Sample Preparation and Surface Sterilization

Individual plant samples were washed in running tap water to remove adherent soil. Sections, 2–3 cm in length, were excised with a flamed scalpel. Root sections were taken just below the soil line in younger plants (14 days or less after emergence) and from 5–10 cm below the soil line in older plants (21 days or more after emergence). Stem sections were taken 1–2 cm above the soil line in younger plants and 10 cm above the soil line in older plants.

All sections were blotted dry with a paper-towel and weighed before processing. Stem samples were surface-disinfested in 20% hydrogen peroxide for 10 min and rinsed four times with sterile 0.02 M potassium phosphate buffer, pH 7.0. Surface-disinfestation parameters for all tissues were optimized prior to experimentation. Root samples were surface-disinfested with 1.05% sodium hypochlorite for 10 min and rinsed four times as previously described. A 0.1 ml aliquot was taken from the final buffer wash of each sample and transferred to a tube of tryptic soy broth to serve as a sterility check. Samples were discarded if growth from the sterililty check occurred within 48 hr. Each sample was triturated with a sterile mortar and pestle in 9.9 ml of the final buffer wash. Serial dilutions were made using phosphate buffer, as previously described, and plated with a spiral plater (Spiral Systems, Inc., Bethesda, MD). Each dilution of every sample was plated on 1 plate each of TSA, R2A and SC.

2.2.4 Growth Conditions, Bacterial Counts and Data Analysis

Agar plates were incubated at 28 °C for 48–72 hr except where noted. Colonies were counted with a laser colony counter (Spiral Systems, Inc., Bethesda, MD) and populations were determined by Bacterial Enumeration software (Spiral Systems, Inc., Bethesda, MD) in colony forming units per ml. Populations were transformed to log 10 colony forming units per gram fresh weight (cfu/g-fw) prior to calculating mean population densities.

2.2.5 Isolation and Preservation of Endophytes

At each sampling date, and for each treatment, one representative of each bacterial colony morphology was transferred to a fresh TSA plate to establish pure cultures. Individual strains were shaker-cultured at room temperature for 18–24 hr in tryptic soy broth. Cultures were then centrifuged at 5000 x g for 7 min at 4 °C. The resulting pellet was resuspended in 2.0 ml TSB amended with 20.0% glycerol and maintained at −80 °C in Nalgene cryovials for later identification by MIS as outlined below.

2.2.6 Strain Identification

Each strain was identified by membrane fatty acid analysis using the Microbial Identification System (11). Strains that could not be identified with a similarity index above 0.100 were considered unidentified.

2.3 Results

2.3.1 Population Dynamics

Bacteria were recovered from surface-disinfested stems and roots of cotton and sweet corn during both growing seasons on all media. Populations from medium R2A and medium SC were significantly greater than populations on TSA ($P=0.0001$). Populations from medium R2A were not significantly different from medium SC ($P=0.0001$). Plate counts from medium R2A were more accurately determined because of less colony overlap and smaller colony size which was due to the low nutritional status of the medium. For these reasons, data are presented from medium R2A.

Total endophytic bacterial (TEB) populations of sweet corn roots and stems (Fig. 1) from the field showed that endophytic bacteria were present at emergence at 10^4 cfu/g-fw for both seasons. TEB populations in corn stems and roots in 1990 remained between 10^4 – 10^6 cfu/g-fw for most of the growing season. These populations increased to 10^8 – 10^{10} cfu/g-fw post-harvest. TEB populations in 1991 were from 10^4 – 10^7 cfu/g-fw for the entire growing season. Although not significant, there was no similar population increase in cotton roots or stems at the end of the 1991 growing season.

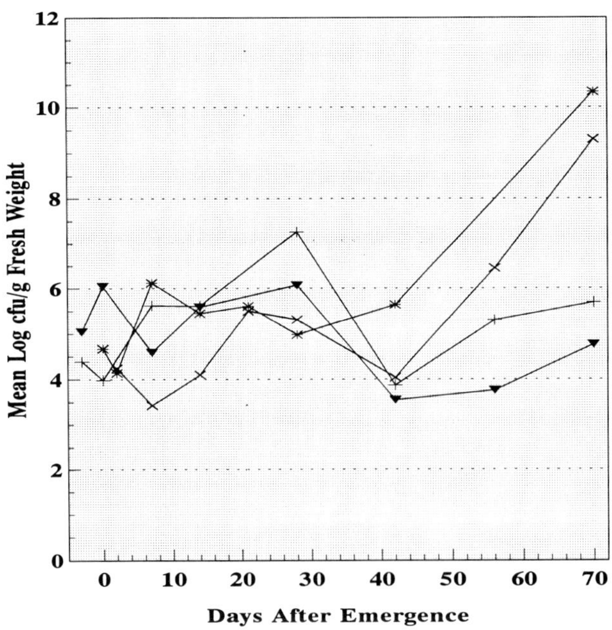

Fig. 1. Population densities of endophytic bacteria from roots (*) and stems (x) of field-grown sweet corn, 1990; and roots (+) and stems (▼) of field-grown sweet corn, 1991.

Endophytic bacteria were present at emergence in cotton roots in 1990 and 1991. In 1990, TEB populations from field-grown cotton roots (Fig. 2) were 10^4 cfu/g-fw for the first week and from 10^5 – 10^8 cfu/g-fw for the rest of the season. In 1991, TEB populations from cotton roots were 10^7 cfu/g-fw during the first week and 10^4 – 10^6 cfu/g-fw for the rest of the season. No cotton stem populations in 1990 were detected at emergence, but bacteria were present 2 days after emergence at 10^3 cfu/g-fw. Cotton stem populations in 1990 remained between 10^4 – 10^6 cfu/g-fw for the rest of the season. In 1991, TEB populations in cotton stems were 10^7 cfu/g-fw at emergence and 10^6 – 10^7 cfu/g-fw for the first week. For the remainder of the season cotton stem populations in 1991 ranged from 10^3 – 10^6 cfu/g-fw.

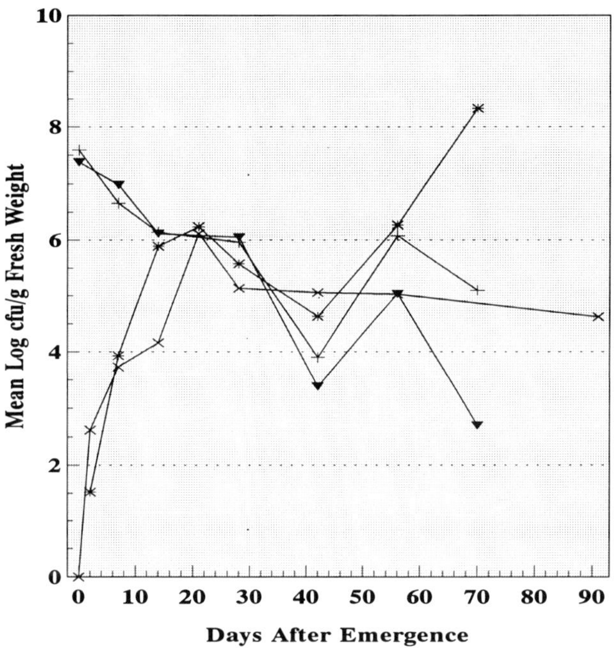

Fig. 2. Population densities of endophytic bacteria from roots (*) and stems (x) of field-grown cotton, 1990; and roots (+) and stems (▼) of field-grown cotton, 1991.

2.3.2 Bacterial Identification

A total of 947 bacterial endophytes were isolated; 313 were from sweet corn roots, 230 from sweet corn stems, 250 from cotton roots and 154 from cotton stems. The endophytic bacteria isolated comprised 34 genera; 31 of these were present in sweet corn and 31 were present in cotton. Twenty five of the 34 genera were Gram-negative taxa. Of the total isolates, 71.4% were Gram-negative, and 25.9% were Gram-positive. Bacteria which were unidentifiable by MIS represented 2.6% of the total. Results of bacterial identification by fatty acid analysis (Table 1) indicated that the diversity of bacteria did not vary between sweet corn and cotton; however, the frequency of occurrence did. The most frequently isolated groups were *Pseudomonas pickettii* and *Pseudomonas solanacearum* from sweet corn roots; *Serratia* spp. from sweet corn stems; *Agrobacterium radiobacter*, *Serratia* spp. and *Staphylococcus* spp. from cotton roots; and *Bacillus megaterium* and *Bacillus pumilus* from cotton stems. *Acinetobacter baumannii*, *Comamonas testosteroni*, and *Cellulomonas* spp. were only isolated from cotton, and *Pantoea agglomerans*, *Flavimonas oryzihabitans*, and *Xanthomonas campestris* pathovars were only isolated from sweet corn. Several taxonomic groups were isolated much more frequently from sweet corn than they were from cotton; these included *Pantoea agglomerans*, *Enterobacter cloacae*, *Pseudomonas cepacia*, *Pseudomonas gladioli*, *Pseudomonas putida*, *Clavibacter* spp., *Klebsiella* spp., and *Kluyvera* spp. There were no taxonomic groups that were isolated much more frequently from cotton than from sweet corn.

In general, bacteria isolated from sweet corn stems were also isolated from sweet corn roots, and vice versa. This was not so in cotton. *Acinetobacter baumannii*, *Bacillus subtilis*, *Arthrobacter* spp., and *Citrobacter* spp. were present in cotton stems but not in cotton roots. There were 13 taxonomic groups present in cotton roots but not in cotton stems, all of which were Gram-negative except for *Microbacterium* spp. isolated at only one sampling date.

Agrobacterium radiobacter was isolated from roots of both crop plants more frequently than from stems of both plants. The group of strains that was unidentifiable came, almost exclusively, from the roots of both crops. All taxonomic groups frequently isolated from stems of both crops were also present in roots of both crops.

2.4 Discussion

Healthy monocotyledonous and dicotyledonous plants were naturally infested with endophytic bacteria at average populations of $10^3 - 10^7$ cfu/g-fw throughout two growing seasons. Endophytes colonized plants early in the season, beginning prior to emergence, based on recovery from seedlings. TEB populations of sweet corn roots and stems, even through germination, generally remained between $10^4 - 10^6$ cfu/g-fw (Fig. 1). Endophytic bacterial populations tended to decrease acropetally, although they do seem to colonize most plant tissues. Root populations were generally slightly greater than stem populations.

The internal tissues of sweet corn and cotton, host a diverse microflora that is similar to common soil bacteria, rhizosphere bacteria, and previously reported endophytic bacteria. Previously reported endophytes have been isolated predominantly from fruits, vegetables, and storage organs of other plant systems and include species of *Bacillus, Agrobacterium, Enterobacter, Erwinia, Flavobacterium, Micrococcus,*

Pseudomonas, Xanthomonas, Citrobacter, and coryneforms including *Curtobacterium, Cellulomonas, Arthrobacter,* and *Clavibacter* (5, 7, 12, 13, 14, 15, 16). The endophytes identified in this study reflect this commonality. However, there are taxonomic groups that have only been previously isolated on one occasion. Lu and Chen (17), among other endophytes already mentioned, identified *Chromobacterium* spp. from cotton infested with *Fusarium*. Gardner et al. (18) identified *Enterobacter sakazakii, Pseudomonas aeruginosa, Serratia liquefaciens, Acinetobacter lwoffii, Yersinia* spp., *Shigella* spp, *Achromobacter* spp., *Providencia* spp., and *Vibrio* spp. from lemon roots. Mundt and Hinkle (6) identified 395 endophytic bacteria from seeds and ovules of 27 different plants and reported several species that had not previously been shown to colonize internal plant tissues. These bacteria, and some of the endophytes that were isolated on only one or two sampling dates in this study, represent what can be called casual opportunistic colonizers of internal plant tissues, e. g., *Acinetobacter, Comamonas, Alcaligenes, Flavimonas,* and *Microbacterium* (Table 1). Since these groups of bacteria are not common soil inhabitants, they probably exist in the environment in association with plants, either with decomposing organic matter, in the rhizosphere, or in the rhizoplane and phylloplane. They may colonize internal plant tissues through natural avenues but are not competitive with other endophytes.

Table 1. Identification and isolation frequency of bacterial endophytes from sweet corn and cotton.

Taxa[1]	Tissue Source Yielding Endophytic Bacteria			
	Sweet Corn		Cotton	
	Root	Stem	Root	Stem
Acinetobacter baumannii				+
Agrobacterium radiobacter	+	+	+	+
Alcaligenes piechaudii		+	+	
Arthrobacter spp.	+	+	+	
Aureobacterium spp.		+	+	+
Bacillus megaterium	+	+	+	+
Bacillus pumilus	+	+	+	+
Bacillus subtilis	+	+	+	
Bacillus thuringiensis	+	+	+	+
Bacillus spp.	+	+	+	+
Cellulomonas spp.		+	+	
Citrobacter spp.	+	+	+	
Clavibacter spp.	+	+	+	+
Comamonas testosteroni			+	
Curtobacterium spp.	+	+	+	+
Enterobacter asburiae	+	+	+	+
Enterobacter cloacae	+	+	+	
Enterobacter taylorae	+	+	+	
Erwinia spp.	+	+	+	+
Escherichia spp.	+	+	+	
Flavimonas oryzihabitans	+			
Flavobacterium spp.	+	+	+	
Hydrogenophaga pseudoflava	+	+	+	+
Klebsiella spp.	+	+	+	+
Kluyvera spp.	+	+	+	

Table 1. (continued).

Taxa[1]	Tissue Source Yielding Endophytic Bacteria			
	Sweet Corn		Cotton	
	Root	Stem	Root	Stem
Methylobacterium spp.	+	+	+	+
Microbacterium spp.	+	+	+	
Micrococcus spp.	+	+	+	+
Ochrobactrum anthropi		+	+	
Pantoea agglomerans	+	+		
Phyllobacterium spp.	+	+	+	+
Pseudomonas cepacia	+	+	+	+
Pseudomonas chlororaphis	+	+	+	+
Pseudomonas gladioli	+	+	+	
Pseudomonas pickettii	+	+	+	+
Pseudomonas putida	+	+	+	
Pseudomonas saccharophila	+	+	+	+
Pseudomonas solanacearum	+	+	+	+
Pseudomonas fluor. spp.	+	+	+	+
Pseudomonas nonfluor. spp.	+	+	+	+
Rhizobium japonicum	+		+	+
Salmonella spp.	+	+	+	+
Serratia spp.	+	+	+	+
Sphingomonas paucimobilis	+	+	+	
Staphylococcus spp.	+	+	+	+
Variovorax paradoxus	+	+	+	
Xanthomonas campestris	+	+	+	+
Xanthomonas maltophilia	+			
Unknown[2]	+	+	+	

[1] Grouped taxa consist of the following species; *Arthrobacter crystallopoietes, A. globiformis, A. mysorens, A. pascens; Aureobacterium barkeri, A. saperdae, A. testaceum; Bacillus amyloliquefaciens, B. cereus, B. coagulans, B. laterosporus, B. lentus, B. licheniformis, B. macerans, B. mycoides, B. pabuli, B. pasteurii, B. polymyxa, B. psychrophilus, B. sphaericus; Cellulomonas cartae, C. cellulans; Clavibacter michiganense* subsp. *insidiosum, C. michiganense* subsp. *nebraskense; Citrobacter diversus, C. freundii; Curtobacterium flaccumfaciens* subsp. *flaccumfaciens, C. flaccumfaciens* subsp. *oortii, C. flaccumfaciens* subsp. *poinsettiae, C. pusillum; Erwinia carnegieana, E. carotovora* subsp. *carotovora, E. herbicola, E. uredovora; Escherichia coli, E. hermannii; Flavobacterium indologenes, F. meningosepticum; Klebsiella planticola, K. pneumoniae* subsp. *ozaenae, K. terrigena; Kluyvera ascorbata, K. cryocrescens; Methylobacterium fujisawaense, M. mesophilicum, M. radiotolerans, M. rhodesianum; Microbacterium imperiale, M. laevaniformans; Micrococcus agilis, M. kristinae, M. luteus, M. lylae, M. roseus, M. varians; Pantoea agglomerans, P. ananas; Phyllobacterium myrsinacearum, P. rubiacearum; Pseudomonas* (fluorescent species) *P. coronafaciens, P. cichorii, P. fluorescens, P. syringae; Pseudomonas* (nonfluorescent species) *P. diminuta, P. marginalis, P. rubrisubalbicans, P. vesicularis; Salmonella bongori, S. choleraesuis* subsp. *arizonae, S. choleraesuis* subsp. *diarizonae, S. choleraesuis* subsp. *houtenae, S. choleraesuis* subsp. *salamae; Serratia marcescens, S. plymuthica, S. proteamaculans* subsp. *proteamaculans; Staphylococcus capitis* subsp. *capitis, S. capitis* subsp. *ureolyticus, S. cohnii, S. epidermidis, S. hominis, S. warneri.*
[2] Bacteria unable to be identified by MIS, 25 total.

Root tissues of both crops generally harbored the same endophytes found in stem tissue, with a few exceptions. The unidentified strains and *A. radiobacter* came almost exclusively from root tissue, suggesting that these microbes are strict colonizers of internal root tissue as opposed to stem tissue. There also were taxa which, although isolated from both stem and root, were more frequently isolated from one over the other, e. g., *B. megaterium* and *B. subtilis* in cotton stems, and *P. cepacia, P. gladioli* and *P. solanacearum* in sweet corn roots. This suggests that strains can adapt to specific plant tissues. The total number of bacteria isolated from each tissue is an indirect measure of tissue diversity, since the bacteria were selected based on unique colonial morphology per treatment and per replicate. The number of endophytes isolated from roots of both crops is greater than that of stems, and the number of endophytes isolated from sweet corn tissues surpasses that isolated from the respective tissues in cotton. These data indicate that internal sweet corn tissues support a more diverse microbial flora than cotton. They also support the hypothesis that bacterial endophytes originate in the rhizosphere and from there proceed into stem tissue.

In order to make valid comparisons of endophytic bacteria with rhizosphere or soil bacteria, surveys of rhizobacteria and soil microbes of the past will have to be re-evaluated. This is due, in part, to the fact that most identification studies were conducted to the genus level only. But re-evaluation is also necessary to compensate for the changing bacterial nomenclature that has taken place over the past 15 years. The *Pseudomonas* genus alone has been fragmented into *Acidovorax, Comamonas, Flavimonas, Hydrogenophaga, Methylobacterium*, and *Sphingomonas*. Plant-associated members of the genus *Corynebacterium* are now in *Aureobacterium, Clavibacter, Curtobacterium*, and *Rathayibacter.*

Endophyte colonization of sweet corn and cotton tissues shown in this study suggests that internal plant habitats are exploited by a wide variety of bacteria. Screening of endophytic bacteria as potential plant growth-promoters and biological control agents can now include representatives from more diverse bacterial taxa, and the list may lengthen as more crops are studied.

2.5 References

1. Samish, Z., R. Etinger-Tulczynska, and M. Bick. Microflora within healthy tomatoes. Appl. Microbiol. *9* (1961) 20–25.
2. Hollis, J.P. Bacteria in healthy potato tissue. Phytopathology *41* (1951) 350–366.
3. Samish, Z., and D. Dimant. Bacterial population in fresh, healthy cucumbers. Food Manuf. *34* (1959) 17–20.
4. Fry, S.M., and R.D. Milholland. Multiplication and translocation of *Xylella fastidiosa* in petioles and stems of grapevine resistant, tolerant, and susceptible to Pierce's disease. Phytopathology *80* (1990) 61–65.
5. Philipson, M.N., and I.D. Blair. Bacteria in clover root tissue. Can. J. Microbiol. *3* (1957) 125–129.
6. Mundt, J.O., and N.F. Hinkle. Bacteria within ovules and seeds. Appl. Environ. Microbiol. *32* (1976) 694–698.
7. Misaghi, I.J., and C.R. Donndelinger. Endophytic bacteria in symptom-free cotton plants. Phytopathology *80* (1990) 808–811.
8. van Peer, R., H.L.M. Punte, L.A. de Weger, and B. Schippers. Characterization of root surface and endorhizosphere pseudomonads in relation to their colonization of roots. Appl. Environ. Microbiol. *56* (1990) 2462–2470.

9. Dimock, M.B., R.M. Beach, and P.S. Carlson. Endophytic bacteria for the delivery of crop protection agents. In: Proceedings of a conference on biotechnology, biological pesticides and novel plant-pest resistance for insect pest management. D.W. Roberts and R.R. Granados (Eds.). Boyce Thompson Institute for Plant Research, Ithaca, New York, 1989, pp. 88–92.
10. Davis, M.J., A.G. Gillaspie Jr., R.W. Harris, and R.H. Lawson. Ratoon stunting disease of sugarcane: Isolation of the causal bacterium. Science *210* (1980) 1365–1367.
11. Sasser, M. Identification of bacteria through fatty acid analysis. In: Methods in phytobacteriology. Z. Klement, K. Rudolph, and D. Sands (Eds.). Akademiai Kiado, Budapest, 1990, pp. 199–204.
12. De Boer, S.H., and R.J. Copeman. Endophytic bacterial flora in *Solanum tuberosum* and its significance in bacterial ring rot diagnosis. Can. J. Plant Sci. 54 (1974) 115–122.
13. Jacobs, M.J., W.M. Bugbee, and D.A. Gabrielson. Enumeration, location, and characterization of endophytic bacteria within sugar beet roots. Can. J. Bot. *63* (1985) 1262–1265.
14. Meneley, J.C., and M.E. Stanghellini. Detection of enteric bacteria within locular tissue of healthy cucumbers. Journal of Food Science *39* (1974) 1267–1268.
15. Samish, Z., R. Etinger-Tulczynska, and M. Bick. The microflora within the tissue of fruits and vegetables. Journal of Food Science *28* (1963) 259–266.
16. Sanford, G.B. The occurrence of bacteria in healthy potato plants and legumes. Scientific Agriculture *28* (1948) 21–25.
17. Lu, S., and Y. Chen. Preliminary studies on the main groups of microorganisms colonizing the vascular system of cotton plant and their population dynamics. Acta Agriculture Universitatis Pekinensis *15* (1989) 326–329.
18. Gardner, J.M., A.W. Feldman, and R.M. Zablatowicz. Identity and behavior of xylem-residing bacteria in rough lemon roots of Florida citrus trees. Appl. Environ. Microbiol. *43* (1982) 1335–1342.

3 Detection of Introduced Bacteria in the Rhizosphere Using Marker Genes and DNA Probes

Maarten H. Ryder, Clive E. Pankhurst, Albert D. Rovira, Raymond L. Correll and Kathy M. Ophel Keller

3.1 Introduction

Micro-organisms are introduced into the soil and rhizosphere to improve plant growth through nitrogen fixation, biological control of disease, formation of mycorrhizae and direct growth stimulation. We need to be able to monitor introduced organisms specifically, to study their fate in the environment and thereby to help explain the success or failure of the inoculation. We also need to monitor the fate of genetically manipulated organisms after they are released into the environment. There is now a wide variety of methods to track microbes in the soil. In addition to the traditional antibiotic resistance and selective plating methods, we can use introduced marker genes, different types of immunological methods and specific DNA and RNA probes. Most methods have been developed for bacteria, but ways of tracking fungi are also being developed.

All detection methods have advantages and disadvantages, and the choice depends on the application. Considerations in selection of a method are: cost, detection limit, degree of specificity, ability to detect non-culturable cells, ease of processing samples and the ability to quantitate results and apply statistics. Another consideration is whether the method allows us to visualise the organism specifically under the light or electron microscope. It is common for researchers to use a combination of two or more techniques to detect a particular organism. For example, spontaneous antibiotic resistant mutants are often used in combination with inserted marker genes or antibody-based detection. The development of new detection methods, based largely on DNA technology and immunology, has greatly increased our ability to follow the survival and distribution of introduced micro-organisms, including genetically-manipulated organisms and the foreign genes that they carry. Combinations of methods can be used for very sensitive and specific detection.

3.2 Methods

The three sections 3.2.1 to 3.2.3 describe detection of introduced bacteria using spontaneous antibiotic resistance, marker genes and DNA probes. Each section contains examples, and discussion of the advantages and disadvantages of the method. In this paper, selectivity refers to lack of background reaction or cross reaction. A highly selective method allows detection without interference from other organisms. Sensitivity refers to detection limit. A very sensitive method allows the enumeration or detection of low populations of an organism.

3.2.1 Spontaneous Antibiotic Resistance

The most common means of detecting specific bacteria has traditionally been to use a selective medium for recovery of a spontaneous antibiotic-resistant derivative of the parent strain. Natural antibiotic resistance can also be used for detection of certain organisms. This method is still often used in conjunction with newer methods and still has unique advantages. Because of its frequent use in conjunction with other means of detection, this technique will be included here. The bacterial strain is marked by selecting a derivative that is resistant to an antibiotic, and the marked strain is detected by plating samples on to solidified agar media containing the antibiotic. Dilution plating, using spread plate, poured plate, or droplet plating methods and the selective medium allows enumeration of microbial populations. Requirements are that the organism be culturable, that there be a low (or non-detectable) natural background of resistance to the antibiotic in the samples being tested, and that the mutation to antibiotic resistance has not significantly impaired the growth and function of the test strain. There are many examples of the use of spontaneous antibiotic resistant derivatives to monitor bacteria in the rhizosphere, including biological control agents such as *Pseudomonas* spp (1, 2, 3) and *Agrobacterium* (4). Most examples are for Gram-negative bacteria, but some Gram-positive organisms (*Bacillus* spp and *Clavibacter* = *Corynebacterium*) have been tracked by this method (5, 6). The most popular antibiotics have been rifampicin and nalidixic acid.

Virtually all genetically manipulated organisms so far released into the environment have been monitored. Antibiotic resistance is useful, especially where the genetic modification itself has not provided a selectable marker, for example in the case of a deletion of genetic material (7). In other cases, the use of a spontaneous antibiotic resistant derivative has made the selective isolation of the genetically-manipulated organism from the environmental sample more straightforward. Rifampicin resistance has been used together with *lacZY* marker genes in pseudomonads (8) and with chromosomally inserted kanamycin resistance in *Erwinia carotovora* (9). Some combinations of soil bacterium and antibiotic resistance may be quite good whereas others may be poor. Acea et al (6) showed that for some soil bacteria but not for others, the survival of antibiotic-resistant mutants in soil compared well to that of the parent strain. Compeau et al (10) emphasized the point that while some rifampicin-resistant mutants of *Pseudomonas* survived as well as the parent strain in soil, others did not.

The unintended alteration of bacterial properties, such as growth rate and competitiveness, that occurs amongst antibiotic-resistant derivatives can be a serious problem. Tests are normally performed to verify that the antibiotic-resistant derivative behaves in the same way as the parent *in vitro* (eg growth rate in complex and minimal media) and also in the rhizosphere. Competition experiments can be useful, and ability to control disease or rate of nitrogen fixation, can also be evaluated. The saturation constant (K_s) of the derivative could also be measured in chemostat culture. This would determine growth at less-than-optimal substrate concentrations, which the organisms would be likely to experience in the environment. There has been some uncertainty about the use of antibiotic resistance in long-term environmental studies. Antibiotic-resistant strains can be difficult to reisolate from the environment. For example there is some evidence of decreased antibiotic resistance in starved cells. Devanas et al (11) showed that *Escherichia coli* strains carrying plasmid-based antibiotic resistance were recovered poorly from soil unless they were first grown on a non-selective medium. Care should therefore be taken when attempting to recover bacteria from an environment where there are stresses such as starvation.

There have been numerous studies of long-term survival of *Rhizobium* strains in soil, but there have been few reports of the long-term stability (months or years) of antibiotic resistance markers in biological control agents. Glandorf et al. (12) demonstrated that rifampicin resistance was a suitable marker for *Pseudomonas fluorescens* in field studies over a period of four months. We do not know enough to make general conclusions. However, we can now cross-check results obtained using antibiotic resistance with those from other methods (see later sections) for a better evaluation of the method. Advantages are that the method is simple, relatively sensitive and rapid. The data can be analysed statistically, and the materials are usually inexpensive. The detection limit can be quite good; as low as 10^3 cfu/gram of sample. The actual limit depends upon the size of sample that can be conveniently processed. Sensitivity can be increased by enrichment and use of Most Probable Number (MPN) enumeration (13). Successful detection by enrichment requires strong selection against the growth of other microbes and a low frequency of new antibiotic resistance mutations in the sample, both during the enrichment phase and during the subsequent selection phase, if performed. Disadvantages are that the method is limited to soils where there is a low or non-detectable background of resistance to the particular antibiotic. A clean background is not always obtainable (14). The use of a double-marked strain can overcome this problem, but this increases the risk of having a strain that does not perform or grow as well as the parent strain.

3.2.2 Marker Genes

The use of marker genes involves the addition of new DNA to an organism so that it can be uniquely identified and distinguished from other microorganisms in the environment into which it is introduced. In this paper, marker genes are defined as genes that are introduced into an organism, either on a plasmid or as an insertion on the chromosome, and which allow the organism to be detected as a result of the expression of the genes. A suitable delivery system is required in order to mark the strain of choice, and the strains must be amenable to basic genetic manipulation. The question of plasmid versus chromosomal insertion is important, because of the possibility of horizontal gene transfer (15). The objectives of the work will determine which marker location is more appropriate. If the aim is to monitor a particular organism rather than a gene, chromosomal markers are preferred. If a plasmid-based marker gene is used, information on the transfer frequency *in vivo* would be desirable before field release.

Improved sensitivity can be gained by enrichment coupled with MPN analysis (13, 16). Both selective plating and nucleic acid probes can be used for detection and enumeration of low populations of organisms carrying marker genes (16). Scanferlato et al (13) monitored a genetically-manipulated strain of *Erwinia carotovora* by MPN analysis after incubating soil with a specific substrate (polypectate), and the antibiotics kanamycin, rifampicin and cycloheximide. A number of different types of marker gene have been used (Table 1).

Tab. 1. Marker Genes for Rhizosphere Bacteria

	Example	Reference
1. New metabolic capacility	lactose utilization (*lacZY*)	8, 17, 18
	catechol dioxygenase (*xylE*)	21, 22, 23
	β-glucuronidase activity (GUS)	20
2. Heavy metal resistance	mercury resistance (*mer*)	24
	arsenite resistance (*ars*AB)	24
3. Bioluminescence	*lux* operon (or part thereof)	25, 26
4. Herbicide resistance	bialaphos resistance (*bar*)	24
5. Transposon-coded antibiotic resistance	Tn*5*, Tn*903*	13, 16

Examples

3.2.2.1 New metabolic capability

Lactose utilization (*lacZY* from *E. coli*) has been introduced into fluorescent pseudomonads as a plasmid-based or a Tn7-mediated chromosomal marker (17, 8). The method has also been used to track the non-fluorescent *Pseudomonas corrugata* this chapter). Other *lacZY*-based methods include: (a) using *lacZ* from *Rhizobium* (18) and (b) a Mu d(*lac*) element to mark *P. aeruginosa* (19). These methods all rely on the use of synthetic β-galactosides (eg 5-chloro-4-bromo-3-indolyl-β-D-galactopyranoside, "X-gal") as substrates to allow ready identification of colonies of the marked strain on selective media. The β-galactosidase activity is constitutive in two cases (17, 19). Similarly, β-glucuronidase from *E. coli* can also be used as a marker for detection of bacteria in the environment (20), with the substrate "X-gluc" in a selective medium.

The *xylE* gene from *Pseudomonas*, which codes for catechol-2,3-dioxygenase, is a convenient marker in that a yellow colour reaction in colonies is used to identify bacteria carrying the gene. This marker has been used to study the survival of *P. putida* in water and in soil (21, 22, 23).

3.2.2.2 Heavy metal resistance

Genes encoding resistance to mercury, organomercuric compounds and to arsenite are available for monitoring bacteria. Resistance to mercury is encoded by the *mer* operon of *Serratia marscescens*. Herrero et al (24) developed methods for marking the chromosome of *Pseudomonas putida* and a range of Gram-negative bacteria with part of the *mer* operon.

3.2.2.3 Bioluminescence

Genes that confer bioluminescence have been isolated from *Vibrio fischeri* (27) and *V. harveyi* (28). Whole or part of the *lux* operon has been used to monitor not only the survival and distribution of introduced bacteria (25, 29, 30, 31, 32), but also the activity of the organism (33, 34). The *lux* operon, or part of the operon, has been used to monitor *Xanthomonas* (25), *Pseudomonas* spp. (29, 31, 32, 33, 35) and *Enterobacter* (30). The insertion of *luxAB* genes from *V. fischeri* to the chromosome of *Pseudomonas fluorescens* strain 10586 did not impose a detrimental load on the organism (35).

3.2.2.4 Herbicide resistance

Resistance to the herbicide bialaphos (a tripeptide, phosphinotricin or "ptt", which is active against bacteria and plants) is encoded by the *bar* gene of *Streptomyces hygroscopicus* (24). Ramos et al (36) used the *ptt* marker together with plasmid-coded p-ethylbenzoate degradation to specifically select a strain of *P. putida.*

3.2.2.5 Transposons carrying antibiotic resistance

Transposons which code for antibiotic resistance have been used frequently to monitor introduced organisms in soil and rhizosphere. The most commonly used transposons are Tn*5* and Tn*903*, both of which encode kanamycin resistance. Bacteria that have been monitored in microcosms include *Azospirillum* (37), *Pseudomonas* spp. (38, 39, 40), *Rhizobium* (16) and *E. coli* (41).

The general advantages of using marker genes are similar to those listed for antibiotic-resistant mutants (3.2.1.1). These are that the methods are simple, relatively inexpensive, quantitative, and statistics can be applied. Table 1 shows that a wide variety of functions is available. Marker genes can offer a good specificity and sensitivity but this is sometimes only achieved by using a second, selectable marker such as spontaneous antibiotic resistance (8). An additional advantage is that the DNA sequences of the marker genes are usually well known and nucleic acid probes can be used to monitor the organism (see 3.2.3). Additional advantages of using bioluminescence genes are that cells can be visualised *in situ* (32), cells that are metabolically active but not culturable on agar media can be detected and the metabolic activity of the organism *in situ* may be assessed (33, 34). The method can be sensitive enough to reveal single cells of the marked strain (32), depending on the level of expression of the *lux* genes.

Transposon-coded antibiotic resistance is convenient to use. Provided that the strain is amenable to insertion of transposons, these markers can be used in the same way as spontaneous antibiotic resistance markers, but with the added advantage that specific DNA probes can be used for detection (42). Increased sensitivity can be gained by enrichment and selective plating (16). Disadvantages are that some markers are not generally applicable: for example *lacZY* can only be used for lac^- organisms. Only culturable cells are recovered and enumerated. Double marking, ie with antibiotic resistance as well as the marker gene is often necessary to achieve good selectivity and sensitivity. A further consideration is that the insertion of genes may decrease the fitness of the organism. For example, the *lacZY* marker genes (constitutive, chromosomal) significantly decreased the survival of the biocontrol agent *P. corrugata* compared to the parent strain in the field over seven months in late spring and summer in South Australia (Figure 1, 43). The reasons for this are

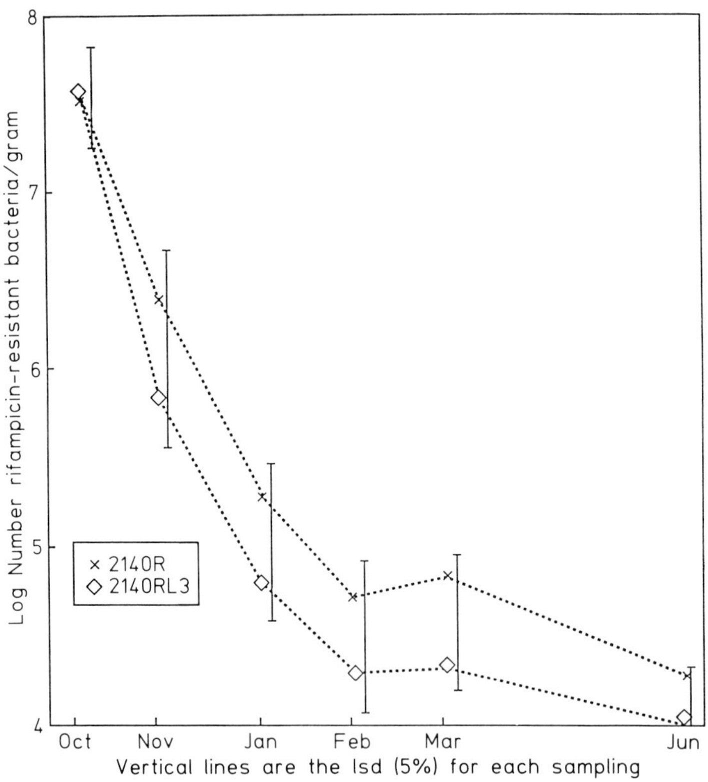

Fig. 1. Population dynamics of *P. corrugata*, with (△) and without (X) *lacZY* genes from *E. coli*, on the roots of field-grown wheat. Wheat (cv Spear) seed was coated with approx. 10^7 bacteria per seed in methyl cellulose, immediately prior to sowing in October 1990. The first two data points are for rhizospheres of actively growing wheat plants. The remaining samples consisted of wheat crowns collected in summer and autumn (1990–91).

not known but could be due to either the constitutive expression of the *lacZ* and *Y* genes (metabolic load) or the position of the insertion in the genome affecting the fitness of the organism by interrupting another cell function.

For transposons carrying antibiotic-resistance genes, there are several disadvantages. Strains carrying this type of marker are not favoured for environmental release because some authorities consider it undesirable to release new antibiotic resistance determinants, especially resistance to medically important antibiotics. Plasmid-based transposons are even less favoured, because plasmids are frequently conjugal or if not, may be mobilized to other bacteria, not necessarily of the same species or genus. As a safeguard it is possible to use disarmed transposons, inserted in the chromosome. However there is still a possibility that resistance could be transferred. It would therefore be prudent to use these markers only for studies conducted in a contained environment.

3.2.3 DNA Probes

Bacteria can be detected in the rhizosphere by the use of specific probes to DNA or RNA sequences. A DNA probe is normally a short DNA sequence that matches and will bind uniquely to DNA of a particular organism or group of organisms, depending on the level of specificity desired. The sequences to which the probe binds in the organism's genome may be naturally occurring and unique in a particular environment. Alternatively, unique sequences can be produced or introduced via genetic manipulation. Within the latter category, probes can be made to detect introduced genes (eg. Tn*7*-based *lacZY* insertion, 44). It is usually possible to derive nucleic acid probes that recognize introduced marker genes because the nature and the sequence of the foreign genes are usually well known. Deletion of DNA generates a junction that may have a unique DNA sequence. Lindow and Panopoulos (7) detected an ice$^-$ derivative of *P. syringae* using a 21-bp probe sequence to the junction generated by deletion of part of the *ice* gene. Synthetic nucleotide sequences can be inserted into the genome and used with specific DNA probes to detect the organism (45). Probes can also be specifically targeted to rRNA (46).

The specificity of the nucleic acid probe needs to be checked thoroughly with samples collected from the environment being studied before it is applied. DNA probes can be used to specifically detect organisms either by:

(1) colony hybridization, i.e. hybridization of the probe to DNA from colonies that have been grown on culture media (47, 48, 49, 50).
(2) direct detection, i.e. hybridization with DNA extracted directly from soil or plant samples (51, 52, 53, 54, 55, 56).

An example of colony hybridization is the detection of *P. fluorescens*, inoculated into soil, after being marked with Tn*5*. Kanamycin was used for selection and the central part of Tn5 was used as a probe (50).

Some examples of direct detection are:

(a) *Bradyrhizobium japonicum*, carrying Tn*5*, inoculated into soil. The marked bacterium was detected using a labelled probe to the *npt*II (kanamycin resistance) gene. Cells were separated from soil, then lysed to extract DNA (52).
(b) Detection of a strain of *Pantoea agglomerans* containing a *nif* plasmid with Tn*5*, in soil. A 3 kb portion of the *nif* sequences was used as the probe (54).
(c) Detection of *P. cepacia* with a plasmid harbouring genes for degradation of 2,4-D and also Tn*1721* after inoculation into soil. DNA was isolated from the soil, and probed with 1 kb from a second plasmid that carries Tn*1721* (53).

Probes for specific detection of an organism can be based on whole chromosomal DNA, a specific insert in a cloning vector, whole plasmid DNA, part of a 16S rRNA sequence or other specific oligonucleotide sequences. The choice of type and size of probe depends on its specificity in the particular example being considered. The degree of specificity needs to be determined for each situation.

The combination of nucleic acid hybridization using fluorochrome-tagged nucleotides with flow cytometry can allow very sensitive detection in water samples, but the method may be difficult to apply to soil and rhizosphere (46). Nucleic acid hybridization can be made quantitative using MPN-DNA hybridization (16, 37, 49). This method was used, together with an enrichment procedure, to quantify *Rhizobium* and *P. putida* in soil (16). The strains were marked with Tn*5* and rifampicin resistance and plasmid pGS9 DNA was used as a probe. A similar technique was used to quantify *Azospirillum* marked with Tn*5* in soil (37), where whole plasmid DNA was not specific and therefore a 30 bp sequence from the *npt*II gene was used as a probe.

The advantages of using DNA probes include high specificity and sensitivity, once a suitable probe sequence has been identified. Direct detection of sequences of DNA from the environment allows measurement of non-culturable as well as culturable cells. With colony hybridization, only culturable cells are detected. DNA hybridization methodology is now well-developed, and for genetically-engineered organisms the method can be used to follow the foreign gene, rather than the genome. In applications such as slot-blot hybridization, many samples can be processed quickly and routine analysis is possible. Disadvantages are firstly that the colony hybridization method allows detection of culturable cells only. When a non-selective medium is used for colony hybridization, its usefulness may be limited by the low frequency of positive colonies. The use of at least a semi-selective medium will be advantageous, as demonstrated by Steffan et al (49). The methods are relatively expensive compared to standard plating techniques. In some applications, the method would not be suited to routine analysis. When monitoring a gene rather than an organism in the environment, a high frequency of gene transfer to other organisms might not be detected. This could be overcome by using the direct detection method in conjunction with colony hybridization.

3.2.4 Detection Limits, Amplification and Enrichment

The typical detection limits for various methods are shown in Table 2. Sensitivity can vary considerably between methods.

Tab. 2. Typical Detection Limits for Methods for Detection of Bacteria

Method	Detection limit (cells/g)	Example, Reference
Selective plating (antibiotic resistance)	10^2–10^3 routine can be 10 to 100	Numerous
DNA probes (soil DNA)	10^4	*B. japonicum*; Holben et al. (52)
Marker genes (often + antibiotic resistance)		
lacZY	25 (can do <10)	*P. fluorescens*; Drahos et al. (8)
*xyl*E	10 cfu/ml water	*P. putida*; Winstanley et al. (23)
luxAB	10^2–10^3	*E. coli*; Rattray et al. (34)
Tn antibiotic resistance	10–100	*R. leguminosarum*, *P. putida* Fredrickson et al. (16)

3.2.4.1 Increased Sensitivity by PCR Amplification

Amplification of the target DNA sequence can be achieved by use of the polymerase chain reaction (PCR). The use of amplification by PCR has been reviewed recently (57) and in this volume by Simonet (Chapter 10, this volume).

3.2.4.2 Enrichment

Enrichment can be used to increase the population of the target micro-organism, when the level is too low for detection by standard techniques. Enrichment can be followed by selective plating, or dot-blot or slot-blot hybridization and when used together with most probable number estimation, populations can be quantified. Examples of enrichment to recover introduced bacteria include: (a) *Rhizobium*, labelled with Tn*5* (16). The organism was recovered from soil using an enrichment medium containing the antibiotics rifampicin and kanamycin and enumeration was by MPN. (b) enrichment of soil that had been inoculated with *P. fluorescens* carrying the Tn*7*::*lac*ZY marker, by applying lactose to the soil, and then using selective plating and hybridization with part of Tn*7* (44) (c) detection of rifampicin-resistant *P. corrugata* with Tn*7*-based chromosomal *lac*ZY, by incubating soil in liquid medium with rifampicin, cycloheximide and lactose (Ryder, unpublished). The presence and level of the marked organism was checked firstly by selective plating on solidified medium with glucose, X-gal, rifampicin and cycloheximide (with MPN), and secondly by MPN and slot blot hybridization using labelled plasmid pMON7117 (17) a probe to *lac*ZY. An average of 7 culturable cells/100 g soil were detected 2 years after field release (Ryder and Ophel-Keller, unpublished).

Typical sensitivities that can be obtained after enrichment or amplification steps are given in Table 3. Enrichment of the target organism by incubating samples in selective media, or amplification of DNA sequences by PCR allow lower detection limits to be reached, and quantification is possible via the use of MPN.

Tab. 3. Detection Limits for Amplification and Enrichment Procedures

Method	Detection limit (cells/g)	Example
Marker genes plus enrichment	1–20 per g (Hg^R,Tc^R) 1 per 2 g (*lux*, Tc^R, Rif^R)	*E. coli*; Devanas et al. (11) *Xanthomonas*; Shaw et al. (58)
Marker genes/Tn plus enrichment & MPN	1–10 per 10g (Tn*903*, Rif^R) 1–10 per 10g (lacZY, Rif^R)	*Erwinia*; Scanferlato et al. (13) *P. corrugata*; Ryder,unpublished
DNA probe, PCR & MPN (total copies of sequence)	1 per g	*P. cepacia*; Steffan & Atlas (59)
DNA probe, (enrichment & MPN; culturable cells)	1–10 per 10g (*lac*ZY, Rif^R)	*P. corrugata*; Ophel Keller, (unpublished)

3.3 Case Study: Tracking *LacZY*-labelled *Pseudomonas corrugata* in the Field

Pseudomonas corrugata strain 2140 was selected originally as a biological control agent for take-all, a root disease of wheat caused by the soil-borne fungus *Gaeumannomyces graminis* var. *graminis* (60). This strain of *P. corrugata* was isolated from long-term wheat field soil from eastern Australia. After initial field studies that demonstrated a good capacity of the biocontrol agent to colonize wheat roots using a rifampicin-resistant derivative (61), the strain was marked with the *lacZY* genes of *E. coli* for further studies of colonization and long-term survival. This was done in association with Monsanto Co., using the Tn7-based transposition method developed by Barry (17). The *lacZY* genes were inserted into the chromosome, under the control of the *iuc* promoter of *E. coli*, and are expressed constitutively in *P. corrugata* (Ryder, unpublished data). The resulting derivative, *P. corrugata* 2140RL3, grew in liquid culture at a rate that was not distinguishable from that of the (rifampicin-resistant) parent strain (Hemming, unpublished). Approval for field release of the genetically-marked *P. corrugata* was granted by the Australian Genetic Manipulation Advisory Committee (GMAC) in July 1990, subject to satisfactory pre-release testing, and with recommendations about the layout and protection of the field site.

3.3.1 Pre-release Testing

The GMAC requested laboratory tests to assess growth and survival, in soil microcosms with and without added lactose, of the genetically-engineered strain which utilizes lactose, and the parent strain which does not. The tests were performed in soil, collected from the intended field release site, and rifampicin-containing medium was used for selection. Results are shown in Figure 2. Where lactose was added, the popu-

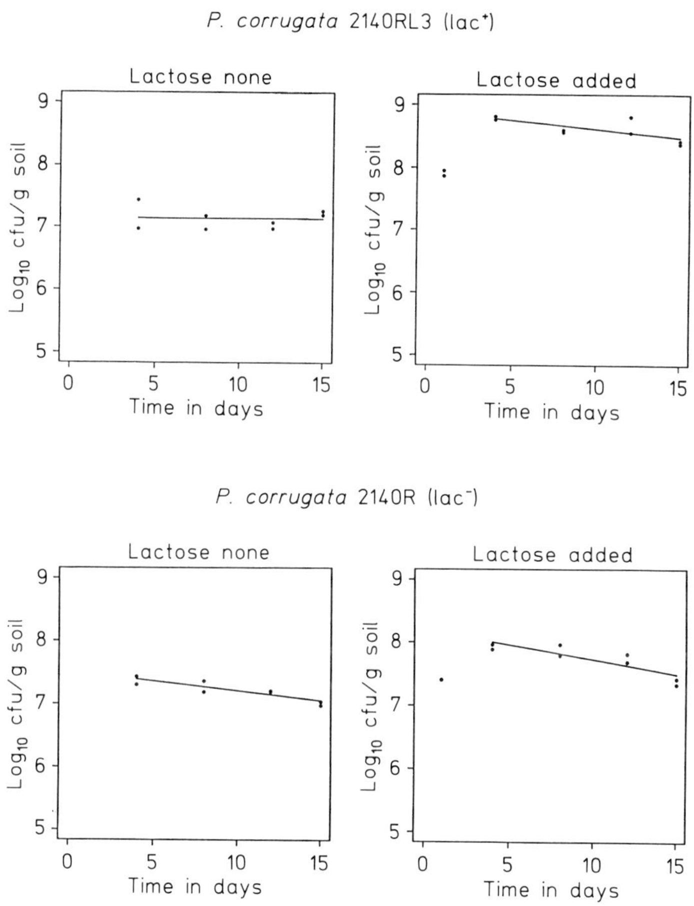

Fig. 2. Pre-release testing of *Pseudomonas corrugata* 2140 (rifampicin-resistant) and its *lacZY*-insertion derivative: population dynamics of the two strains in soil microcosms with and without added lactose. Bacteria (mid-log) were washed in phosphate buffer and added to soil (a sandy loam, collected from Roseworthy, South Australia) at 4 x 10^7 cfu/g dry soil. Where lactose was added, at 1 mg carbon/g dry soil, the soil was also amended with N (as ammonium nitrate) and P (as potassium phosphate). The soils were incubated in the dark at 15 °C, at constant moisture content (14.4% w/w). Duplicate samples were taken at each time, and populations were determined by dilution plating on nutrient agar containing rifampicin (100 mg per litre) and cycloheximide (75 mg per litre).

lations of introduced bacteria reached a maximum at 1 to 4 days after inoculation. Numbers then decreased, in most cases with a measurable half-life, and where it was not possible to determine a half-life, the populations were at the low end of the data set (Ryder and Correll, unpublished). Where no lactose was added, populations of the introduced bacteria were lower than in the lactose-treated soil, and declined in most cases, over the 15 days of the experiment.

3.3.2 Field Release

Final approval for field release was given by the GMAC in October, 1990, 6 months after the initial application was submitted. Previous field releases of soil pseudomonads that were marked with *lacZY* by the same method had been performed in the USA by Monsanto and Clemson University (62). These precedents appear to have assisted the application for field release in Australia. For the field test in South Australia, the GMAC made specific recommendations in addition to the general guidelines laid down for field release in Australia. These recommendations were that: (a) the site be enclosed by a fence, and covered with bird-proof mesh, to prevent access by birds, animals and unauthorized people; (b) the plot areas be surrounded by a ditch and bund to prevent surface water run-off; (c) staff working at the site adhere to a number of safety requirements.

The trial site measured 20 x 20 metres. The three treatments, replicated 6 times, were: wheat sown with a coating of either the parent strain or the engineered strain, applied in methyl cellulose, or methyl cellulose only. In the second year (1991), all plots were sub-divided into three and wheat, peas and rye-grass were sown on the sub-plots, without re-inoculation. In the third season (1992), the plots were left fallow and soil samples were taken, to determine long-term survival.

Results show that the populations of both strains of both parent and engineered bacteria declined quickly in the rhizosphere over the first three months (43, Figure 1). No engineered bacteria were detected on wheat ears at harvest. The bacteria remained detectable on dry wheat crowns over the first summer, and on the roots of all three crops that were sown in the second season. The maximum natural spread of the bacteria in the first season was 20 cm away from the plot area. This dispersal was probably caused by either lateral growth of colonized roots, or by water flow because the site was on slight slope.

In the third season (fallow), the *lacZY*-marked bacteria were no longer detectable by conventional plating techniques. However, an enrichment procedure in which soil samples were incubated with liquid medium containing lactose, rifampicin and cycloheximide, enabled detection and enumeration by MPN. The mean population of *P. corrugata* in field soil from the inoculated plots two years after inoculation was 7 culturable cells/100 grams of soil (10-gram soil samples, 5-replicate MPN). This illustrates the high sensitivity of enrichment procedures. The low level of culturable cells of *P. corrugata* reisolated from the field by enrichment may not be biologically significant. However, there could be situations where the population might increase. For example, planting another crop would give an input of carbon into the soil which could in turn lead to a rise in the population of the marked bacterium.

Risk assessment for introduced bacteria, including genetically-manipulated bacteria, requires two types of information. Firstly, the capacity of the organism to spread and persist in the environment. Information of this type can be gained from the type of field trial described here, where long-term survival and spread from the point of inoculation is measured. A second aspect of risk assessment is the potential for an introduced rhizosphere bacterium to harm "non-target" crop and native plants and the soil biota. A method of assessing the latter risks was proposed by Ryder and Correll (43).

3.4 The Ecological Fitness of Genetically-Engineered Bacteria

Terms that are used to describe the possible deleterious effects of genetic manipulation on soil bacteria include (a) the "metabolic load" of carrying added genetic material and the expression of foreign genes; (b) loss of competitive ability; and (c) decreased fitness.

3.4.1 Metabolic Load

The maintenance and/or expression of introduced genes could cause a drain on the cell's metabolism such that growth and/or survival in the soil environment are decreased relative to the parent strain. The size of the metabolic load depends on the amount and level of expression of the extra DNA. One might question whether the expression of one or two extra genes among the >5,000 in the bacterial genome could affect survival, but it might occur if expression of the foreign gene(s) were at a high level and not regulated, or if the genes were borne on a plasmid present in high copy number. An example of a deleterious effect which is possibly due to metabolic load, is the decreased biological control activity of an antibiotic-overproducing mutant of an *Agrobacterium* strain (63). The mutation resulted in a five-fold increase in copy number of the 48 kb plasmid that encodes agrocin production. This then impaired the cell's ability to perform in the environment.

3.4.2 Reduced Fitness

The terms "loss of competitive ability" and "reduced fitness" appear to refer to the same process. Genetic manipulation could result in reduced fitness via a number of possible mechanisms. Firstly, where DNA is added to the genome, increased metabolic load could be the reason. However, decreased fitness or competitive ability could also be caused by either DNA deletion or by mutation resulting from an insertion. Thus, cell functions involved in either growth or survival could be impaired. Taking the *lacZY*-marking of pseudomonads as an example, Drahos et al (64) reported no difference in long-term survival in the field when comparing the engineered strain to the parent strain of *P. fluorescens*. On the other hand, reduced survival of the marked strain compared to the parent strain of *P. corrugata* in the field was reported from Australia (Figure 1, 43). This could have been due to the metabolic load of constitutive expression of the introduced genes, or to a mutation occurring with the insertion of the Tn7-based element into the chromosome.

Usually, a bacterial strain that has been marked with foreign gene(s) is tested and a derivative which most closely resembles the parent strain is selected. This reduces the chance of using a marked strain that is impaired in growth and survival for field tests. The possible loss of competitive ability is a familiar concept in the process of selecting mutants resistant to antibiotics such as rifampicin. The same principles can be applied to the selection of a suitable marked strain for both of these methods. Tests that are commonly performed include measuring specific growth rate in culture, and competition with the parent strain in liquid culture or in sterile soil. Competition in a chemostat under nutrient-limited conditions could also be added to this list. Small differences in fitness will become amplified with time, so that decreased long-term survival of an engineered strain may not necessarily be predicted from microcosm studies.

Questions still surround the role of soil microcosms in pre-release testing procedure. For example, how closely should a microcosm be "matched" to the field environment? Are the microcosms to be taken from the field intact, or repacked in the laboratory? Do we need to ensure variation in soil conditions such as water potential or temperature, and do we include plants? Soil that is kept at constant moisture and temperature, without plants, may be so different from field soil as to preclude useful prediction. Another consideration is the use of natural soil, partially- or fully-sterilized soil. This determines whether there is a "full spectrum" of microbial competition. The importance of other trophic levels of the food chain, especially predators of bacteria such as amoebae (35, 65) and microbivorous nematodes (66, Bird and Ryder, unpublished) should also be noted. The duration of experiments is important: if the difference in survival between parent and derivative strain is small, it may be revealed only in the long term (eg months). The choice of microcosm will depend on the type of information sought and the nature the organism to be studied. Validation of microcosms by comparison with field results has been reported recently by Bolton et al, (67, 68) and Pedersen (69).

3.5 Conclusions

The diversity of methods available for tracking microorganisms that are introduced into the environment has increased greatly since the introduction of recombinant DNA technology and improved immunological methods. For genetically-engineered bacteria, the choice of methods for detection is greater than for unmodified bacteria.

However, there is potential for the development of isolate-specific nucleic acid probes to ribosomal RNA sequences that could be applied to non-engineered organisms.

The choice of method will depend on the nature of the application, the type of target bacterium, the background interference from other organisms, and the availability and ease of use of genetic markers, DNA probes or antibodies. Other considerations will be the cost, and the number of samples that will be processed routinely. If the organism cannot be easily manipulated genetically, then spontaneous antibiotic resistance or antibodies may be the most appropriate. The purpose of the monitoring program will determine the choice of method. Different techniques will allow us to specifically detect all cells, all intact DNA sequences, culturable cells, or live cells. Some methods allow measurement of gene expression, or the metabolic state of the organism. Marker genes have proven to be extremely useful for tracking bacteria in the soil environment. However there are still unresolved questions on the possible "side-effects" of introducing genes, particularly on growth rate and long-term survival in the environment.

In many cases, two or more detection methods have been combined to good effect. It has been useful to combine selective and diagnostic probes, or marker genes, with other detection methods. There are many examples: transposon-encoded antibiotic resistance + bioluminescence (58); Tn-based antibiotic resistance + DNA probe (42, 70); marker gene + spontaneous antibiotic resistance (12, 36). The advantages of combining methods are (a) greater certainty for the unambiguous detection of the target strain; (b) greater ease of detection; (c) greater sensitivity (lower detection limit).

The need to monitor the fate of genetically-engineered micro-organisms in the environment is clear. Fortunately, the engineered organisms can usually be tracked readily, because the alteration can be detected by using nucleic acid probes. The ability to track both genetically-engineered and non-modified bacteria is increasingly necessary for both ecological studies of the their fate after release, and for the accumulation of data that can be used in risk assessment. The move towards uniform regulations across international boundaries, which is assisted by the efforts of the OECD, will aid the prudent use of genetically-engineered microorganisms in the environment.

Acknowledgments

The authors would like to thank J. A. van Elsas, P. G. Hartel, J. E. Hollebone, C. Jacobsen, K. Killham, J. W. Kloepper, and S. E. Lindow for providing materials prior to publication. The financial support of Monsanto Australia is gratefully acknowledged. M. A. Borrett, CSIRO, provided excellent technical assistance for the experimental work reported here.

3.6 References

1. Davies, K. G., and R. Whitbread. Factors affecting the colonisation of a root system by fluorescent pseudomonads: The effects of water, temperature and soil microflora. Pl. Soil *116* (1989) 247–256.
2. Dupler, M., and R. Baker. Survival of *Pseudomonas putida*, a biological control agent, in soil. Phytopathology *74* (1984) 195–200.
3. Weller, D. M. Colonization of wheat roots by a fluorescent pseudomonad suppressive to take-all. Phyopathol. *73* (1983) 1548–1553.
4. Macrae, S., J. A. Thomson, and J. van Staden. Colonization of tomato plants by two agrocin-producing strains of *Agrobacterium tumefaciens*. Appl. Env. Microbiol. *54* (1988) 3133–3137.
5. Juhnke, M. E., D. E. Mathre, and D. C. Sands. Identification and characterization of rhizosphere-competent bacteria of wheat. Appl. Env. Microbiol. *53* (1987) 2793–2799.
6. Acea, M. J., C. N. Moore, and M. Alexander. Survival and growth of bacteria introduced into soil. Soil Biol. Biochem. *20* (1988) 509–515.
7. Lindow, S. E., and N. J. Panopoulos. Field tests of recombinant ice$^-$ *Pseudomonas syringae* for biological frost control in potato. In: M. Sussman, C. Collins, F. Skinner and D. Stewart-Tull (ed.). Release of Genetically-engineered Micro-organisms. Academic Press, San Diego, 1988, p. 121–138.
8. Drahos, D. J., B. C. Hemming, and S. McPherson. Tracking recombinant organisms in the environment: β-galactosidase as a selectable non-antibiotic marker for fluorescent pseudomonads. Bio/Technology *4* (1986) 439–444.
9. Scanferlato, V. S., D. R. Orvos, J. Cairns, Jr., and G. H. Lacy. Genetically engineered *Erwinia carotovora* in aquatic microcosms: survival and effects on functional groups of indigenous bacteria. Appl. Env. Microbiol. *55* (1989) 1477–1482.
10. Compeau, G., B. J. Al-achi, E. Platsouka, and S. B. Levy. Survival of rifampin-resistant mutants of *Pseudomonas fluorescens* and *Pseudomonas putida* in soil systems. Appl. Env. Microbiol. *54* (1988) 2432–2438.
11. Devanas, M. A., D. Rafaoli-Eshkol, and G. Stotzky. Survival of plasmid-containing strains of *Escherichia coli* in soil: effect of plasmid size and nutrients on survival of hosts and maintenance of plasmids. Curr. Microbiol. *13* (1986) 269–277.
12. Glandorf, D. C. M., I. Brand, P. A. H. M. Bakker, and B. Schippers. Stability of rifampicin resistance as a marker for root colonization studies of *Pseudomonas putida* in the field. Pl. Soil *147* (1992) 135–142.
13. Scanferlato, V. S., D. R. Orvos, G. H. Lacy, and J. Cairns, Jr. Enumerating low densities of genetically engineered *Erwinia carotovora* in soil. Lett. Appl. Microbiol. *10* (1990) 55–59.
14. Schmidt, F. R. J., J. Rosien, and A. Brokamp. The role of soil bacteria in risk assessment analysis, In: J. C. Fry and M. J. Day (ed.), Bacterial Genetics in Natural Environments. Chapman and Hall, London, (1990) p. 207–215.
15. Smit, E., J. D. van Elsas and J. A. van Veen. Risks associated with the application of genetically modified microorganisms in terrestrial ecosystems. FEMS Microbiol. Rev. *88* (1992) 263–278.
16. Fredrickson, J. K., D. F. Bezdicek, F. J. Brockman, and S. W. Li. Enumeration of Tn*5* mutant bacteria in soil by using a most-probable-number-DNA hybridization procedure and antibiotic resistance. Appl. Env. Microbiol. *54* (1988) 446–453.
17. Barry, G. F. A broad-host-range shuttle system for gene insertion into the chromosomes of Gram-negative bacteria. Gene *71* (1988) 75–84.
18. O'Gara F., B. Boesten, and S. Fanning. The development and exploitation of "marker genes" suitable for risk evaluation studies on the release of genetically engineered microorganisms in soil, In W. Klingmüller (ed.), Risk Assessment for Deliberate Releases. Springer, Berlin, 1988. p. 50–64.
19. Höfte, M., M. Mergeay, and W. Verstraete. Marking the rhizopseudomonas strain 7NSK2 with a Mu d(*lac*) element for ecological studies. Appl. Env. Microbiol. *56* (1990) 1046–1052.
20. Jefferson, R. A. The GUS gene reporter system. Nature *342* (1989) 837–838.
21. MacNaughton, S. J., D. A. Rose, and A. G. O'Donnell. Persistence of a *xylE* marker gene in *Pseudomonas putida* introduced into soils of differing texture. J. Gen. Microbiol. *138* (1992) 667–673.

22. Winstanley, C., J. A. W. Morgan, R. W. Pickup, J. G. Jones, and J. R. Saunders. Differential regulation of lambda p_L and p_R promoters by a *c*I repressor in a broad-host-range thermoregulated plasmid marker system. Appl. Env. Microbiol. *55* (1989) 771-777.
23. Winstanley, C., J. A. W. Morgan, R. W. Pickup, and J. R. Saunders. Use of a *xyl*E marker gene to monitor survival of recombinant *Pseudomonas putida* populations in lake water by culture on non-selective media. Appl. Env. Microbiol. *57* (1991) 1905-1913.
24. Herrero, M., V. de Lorenzo, and K. N. Timmis. Transposon vectors containing non-antibiotic resistance selection markers for cloning and stable chromosomal insertion of foreign genes in Gram-negative bacteria. J. Bacteriol. *172* (1990) 6557-6567.
25. Shaw, J. J., F. Dane, D. Geiger, and J. W. Kloepper. Use of bioluminescence for detection of genetically engineered microorganisms released into the environment. Appl. Env. Microbiol. *58* (1992) 267-273.
26. Shaw, J. J., and C. I. Kado. Development of a *Vibrio* bioluminescence gene-set to monitor phytopathogenic bacteria during the ongoing disease process in a non-disruptive manner. Bio/Technology *4* (1986) 560-564.
27. Engebrecht, J., K. Nealson, and M. Silverman. Bacterial bioluminescence: isolation and genetic analysis of functions from *Vibrio fischeri*, Cell *32* (1983) 773-781.
28. Karp, M. Expression of bacterial luciferase genes from *Vibrio harveyi* in *Bacillus subtilis* and in *Escherichia coli*. Biochim. Biophys. Acta. *1007* (1989) 84-89.
29. de Weger, L. A., P. Dunbar, W. F. Mahafee, E. J. J. Lugtenberg, and G. S. Sayler. Use of bioluminescence markers to detect *Pseudomonas* spp. in the rhizosphere. Appl. Env. Microbiol. *57* (1991) 3641-3644.
30. Fravel, D. R., R. D. Lumsden, and D. P. Roberts. *In situ* visualization of the biocontrol rhizobacterium *Enterobacter cloacae* with bioluminescence. Pl. Soil *125* (1990) 233-238.
31. King, J. M. H., P. M. DiGrazia, B. Applegate, R. Burlage, J. Sanseverino, P. Dunbar, F. Larimer, and G. S. Sayler. Rapid, sensitive bioluminescent reporter technology for naphthalene exposure and biodegradation. Science *249* (1990) 778-781.
32. Silcock, D. J., R. N. Waterhouse, L. A. Glover, J. I. Prosser, and K. Killham. Detection of a single genetically modified bacterial cell in soil by using charge coupled device-enhanced microscopy. Appl. Env. Microbiol. *58* (1992) 2444-2448.
33. Meikle, A., K. Killham, J. I. Prosser, and L. A. Glover. Luminometric measurement of population activity of genetically modified *Pseudomonas fluorescens* in the soil. FEMS Microbiol. Lett. *99* (1992) 217-220.
34. Rattray, E. A. S., J. I. Prosser, L. A. Glover, and K. Killham. Matric potential in relation to survival and activity of a genetically modified microbial inoculum in soil. Soil Biol. Biochem. *24* (1992) 421-425.
35. Wright, D. A., K. Killham, L. A. Glover, and J. I. Prosser. The effect of location in soil on protozoal grazing of a genetically modified bacterial inoculum. Geoderma *56* (1993) 633-640.
36. Ramos, J. L., E. Duque, and M-I. Ramos-Gonzalez. Survival in soils of an herbicide-resistant *Pseudomonas putida* strain bearing a recombinant TOL plasmid. Appl. Env. Microbiol. *57* (1991) 260-266.
37. Bentjen, S. A., J. K. Fredrickson, P. van Voris and S. W. Li. Intact soil-core microcosms for evaluating the fate and ecological impact of the release of genetically engineered organisms. Appl. Env. Microbiol. *55* (1989) 198-202.
38. Hartel, P. G., J. W. Williamson, and M. A. Schell. Growth of genetically altered *Pseudomonas solanacearum* in soil and rhizosphere. Soil Sci. Soc. Amer. J. *54* (1990) 1021-1025.
39. Trevors, J. T., J. D. van Elsas, L. S. van Overbeek, and M-E. Starodub. Transport of a genetically engineered *Pseudomonas fluorescens* strain through a soil microcosm. Appl. Env. Microbiol. *56* (1990) 401-408 .
40. van Elsas, J. D., J. T. Trevors, and L. S. van Overbeek. Influence of soil properties on the vertical movement of genetically-marked *Pseudomonas fluorescens* through large soil microcosms. Biol. Fertil. Soils *10* (1991) 249-255.
41. Recorbet, G., A. Givaudan, C. Steinberg, R. Bally, P. Normand, and G. Faurie. Tn*5* to assess soil fate of genetically marked bacteria: screening for aminoglycoside-resistance advantage and labelling specificity. FEMS Microbiol. Ecol. *86* (1992) 187-194.
42. Pillai, S. D., K. L. Josephson, R. L. Bailey, C. P. Gerba, and I. L. Pepper. Rapid method for pro-

cessing soil samples for polymerase chain reaction amplification of specific gene sequences. Appl. Env. Microbiol. *57* (1991) 2283–2286.

43. Ryder, M. H., and R. L. Correll. Assessing the potential benefits and risks of introducing natural and genetically-manipulated bacteria for control of soil-borne root diseases. In H.M.T. Hokkanen and J. M. Lynch (ed.), Biological control: benefits and risks. Cambridge University Press, (in press)
44. Kluepfel, D. A., and D. W. Tonkyn. The ecology of genetically altered bacteria in the rhizosphere. In: E. C. Tjamos, G. C. Papavizas and R. J. and Cook (ed.), Biological control of plant diseases. Plenum Press, NY. 1992, p. 407–413.
45. Amici, A., M. Bazzicalupo, E. Gallori, and F. Rollo. Monitoring a genetically engineered bacterium in a freshwater environment by rapid enzymatic amplification of a synthetic DNA "number-plate?". Appl. Microbiol. Biotechnol. *36* (1991) 222–227.
46. Amann, R. I., B. J. Binder, R. J. Olson, S. W. Chisholm, R. Devereux, and D. A. Stahl. Combination of 16S rRNA-targeted oligonucleotide probes with flow cytometry for analyzing mixed microbial populations. Appl. Env. Microbiol. *56* (1990) 1919–1925.
47. Jain, R. K., G. S. Sayler, J. T. Wilson, L. Houston, and D. Pacia. Maintenance and stability of introduced genotypes in groundwater aquifer material. Appl. Env. Microbiol. *53* (1987) 996–1002.
48. Sayler, G. S., C. Harris, C. Pettigrew, D. Pacia, A. Breen, and K. M. Sirotkin. Evaluating the maintenance and effects of genetically engineered microorganisms. Dev. Ind. Microbiol. *27* (1987) 135–149.
49. Steffan, R. J., A. Breen, R. M. Atlas, and G. S. Sayler. Application of gene probe methods for monitoring specific microbial populations in freshwater ecosystems. Can. J. Microbiol. *35* (1989) 681–685.
50. van Elsas, J. D., J. T. Trevors, D. Jain, A. Wolters, C. E. Heijnen, and L. S. van Overbeek. Survival of, and root colonization by, alginate-encapsulated *Pseudomonas fluorescens* cells following introduction into soil. Biol. Fertil. Soils *14* (1992) 14–22.
51. Chaudhry, G. R., G. A. Toranzos, and A. R. Bhatti. Novel method for monitoring genetically engineered microorganisms in the environment. Appl. Env. Microbiol. *55* (1989) 1301–1304.
52. Holben, W. E., J. K. Jansson, B. K. Chelm, and J. M. Tiedje. DNA probe method for the detection of specific microorganisms in the soil bacterial community. Appl. Env. Microbiol. *54* (1988) 703–711.
53. Jacobsen, C. S., and O. F. Rasmussen. Development and application of a new method to extract bacterial DNA from soil based on separation of bacteria from soil with cation-exchange resin. Appl. Env. Microbiol. *58* (1992) 2458–2462.
54. Selenska, S., and W. Klingmüller. Direct detection of *nif*-gene sequences of *Enterobacter agglomerans* in soil. FEMS Microbiol. Lett. *80* (1991) 243–246.
55. Simonet, P., N. T. Lee, E. T. Du Cros, and R. Bardin. Identification of *Frankia* strains by direct DNA hybridization of crushed nodules. Appl. Env. Microbiol. *54* (1988) 2500–2503.
56. Smalla, K., N. Cresswell, L. C. Mendonca-Hagler, A. Wolters, and J. D. van Elsas. Rapid DNA extraction protocol from soil for polymerase chain reaction-mediated amplification. J. Appl. Bacteriol., 1993, in press .
57. Steffan, R. J., and R. M. Atlas. Polymerase chain reaction: applications in environmental microbiology. Ann. Rev. Microbiol. *45* (1991) 137–161.
58. Shaw, J. J., L.G. Settles, and C. I. Kado. Transposon Tn*4431* mutagenesis of *Xanthomonas campestris* pv. *campestris*: Characterization of a nonpathogenic mutant and cloning of a locus for pathogenicity. Mol. Plant-Microbe Interactions *1* (1988) 39–45.
59. Steffan, R. J., and R. M. Atlas. DNA amplification to enhance detection of genetically engineered bacteria in environmental samples. Appl. Env. Microbiol. *54* (1988) 2185–2191.
60. Ryder, M. H., and A. D. Rovira. Biological control of take-all of glasshouse-grown wheat using strains of *Pseudomonas corrugata* isolated from wheat field soil. Soil Biol. Biochem. *25* (1993) 311–320.
61. Ryder, M. H., and M. A. Borrett. Root colonization by non-fluorescent pseudomonads used for the control of wheat take-all. In: C. Keel, B. Koller and G. Défago (ed.), Plant Growth-Promoting Rhizobacteria – Progress and Prospects. Int'l Org. for Biological and Integrated Control of Noxious Animals and Plants. IOBC/WPRS Bulletin XIV/8. 1991, p. 123–130.

62. Kluepfel, D. A., E. L. Kline, H. D. Skipper, D. J. Drahos, G. F. Barry, B. C. Hemming, D. T. Gooden, T. A. Hughes, and E. J. Brandt. The release and tracking of genetically engineered bacteria in the environment. Phytopathol. *81* (1991) 348-352.
63. Shim, J-S., S. K. Farrand, and A. Kerr. Biological control of crown gall: construction and testing of new biocontrol agents. Phytopathol. *77* (1987) 463-466.
64. Drahos, D. J., G. F. Barry, B. C. Hemming, E. J. Brandt, E. L. Kline and D. A. Kluepfel. Spread and survival of genetically marked bacteria in soil. In: J. C. Fry and M. J. Day (ed.), Release of Genetically-Engineered and Other Microorganisms. Cambridge University Press (UK). 1992, p. 147-159.
65. Heijnen, C. E. and J. A. van Veen. A determination of protective microhabitats for bacteria introduced into soil. FEMS Microbiol. Ecol. 85 (1991) 73-80.
66. Griffiths, B. S. A comparison of microbial-feeding nematodes and protozoa in the rhizosphere of different plants. Biol. Fertil. Soils *9* (1990) 83-88.
67. Bolton, H. Jr., J. K. Fredrickson, S. A. Bentjen, D. J. Workman, S. W. Li, and J. M. Thomas. Field calibration of soil-core microcosms: fate of a genetically altered rhizobacterium. Microb. Ecol. *21* (1991) 163-173.
68. Bolton, H. Jr., J. K. Fredrickson, J. M. Thomas, S. W. Li, D. J. Workman, S. A. Bentjen, and J. L. Smith. Field calibration of soil-core microcosms: ecosystem structural and functional comparisons. Microb. Ecol. *21* (1991) 175-189.
69. Pedersen, J. C. Survival of *Enterobacter cloacae*: field validation of a soil/plant microcosm. Microb. Releases *1* (1992) 87-93.
70. Pillai, S. D., and I. L. Pepper. Transposon Tn*5* as an identifiable marker in rhizobia: survival and genetic stability of Tn*5* mutant bean rhizobia under temperature stressed conditions in desert soils. Microb. Ecol. *21* (1991) 21-33.

4 Impact of GEMMOs on Rhizosphere Population Dynamics

J. M. Lynch, F. A. A. M de Leij, J. M. Whipps and M. J. Bailey

4.1 Introduction

Whereas there has been much interest in recent years in the use of reporter genes to track bacteria which are introduced into the rhizosphere, there have been few reports of the impact of genetically engineered or modified micro-organisms (GEMMOs) on the population dynamics of microbial indigenous communities.

The first consideration is whether genes are exchanged. There have been several studies published (for reviews see Fry et al, (1); Stewart-Tull et al, (2), demonstrating plasmid exchange between bacteria in microcosms. Unlike the natural environment many of those systems studied have not been thermodynamically open. To achieve this it is necessary to use continuous-flow column reactors. One such system has been used to study intrageneric gene transfer between the common rhizosphere bacteria *Enterobacter cloacae* and *Pseudomonas cepacia*, each carrying the transmissible plasmid R383 :: Tn1721 (3). Genes encoding for resistance to tetracycline and trimethoprim were present on this plasmid. Plasmid transfer rates to the same species as recipients, chromosomally resistant to nalidixic acid, were higher in columns containing vermiculite or sterile soil and supplied with nutrient solution than those in sterile and non-sterile soil columns without an input of nutrient solution. In this respect, the nutrient input is potentially mimicking the supply of nutrients from rhizodeposition; it would therefore be expected that exchange frequencies in the rhizosphere would be greater than those in the bulk soil.

This event has recently been demonstrated *in situ*. Lilley et al (4) described the transfer of a conjugative mercury resistance plasmid (pQBR11) between indigenous pseudomonad isolates introduced to the rhizosphere of sugar beet plants in the field. The study plasmid had been exogenously isolated from the microflora associated with the sugar beet rhizosphere and the donor and recipient bacteria were found to be numerically abundant in the same habitat. By monitoring the transfer of pQBR11 from the nalidixic acid donor (*Pseudomonas marginalis* 376N) to a rifampicin resistant recipient (*P. aureofaciens* 381R) the prediction that the rhizosphere stimulates conjugation was confirmed. The frequency of plasmid transfer was found to be greater in those samples collected at the root surface than those from the rhizosphere soil further away from the plant. The *in situ* transfer frequencies were significantly greater than those determined in non-rhizosphere soil amended with nutrient broth. In soils unamended with nutrients no transfer was detected.

The only illustration of natural gene exchange in the field concerns *Agrobacterium*. The wild-type *Agrobacterium radiobacter* strain 84 which carries a plasmid pAg396 encoding for the production of the antibiotic agrocin 84 is an effective agent for the control of Crown-Gall disease of fruit trees caused by *Agrobacterium tumefaciens* (5). The pathogen carries the Ti-plasmid which induces tumours in the plant root which contains plant opines. Strain 84 kills the pathogen because agrocin 84 is actively transported into the pathogen via a permease which also transfers other opines into its own cells. The opines provide a food base to the pathogen, and thus food base and toxin are taken up at the same time. However the genes on plasmid pAg396 controlling agrocin 84 production (and resistance) can be transferred from the antagonist to a pathogenic recipient in the presence of the opine nopaline. This resulted in some situations where the biocontrol agent failed. However, an antagonist strain was engineered with a deficient plasmid-transfer system (6). This strain functioned satisfactorily in the field and the "disarmed wild-type" which effectively became a GEMMO, was cleared for large-scale use by the relevant Australian authorities.

Clearly both the above examples illustrate that genes carried on transmissible plasmids can be transferred to other members of the microbial community, and it appears that regulatory authorities will be concerned about functional genes carried on transmissible plasmids unless the plasmids are disarmed. To reduce the chance of gene transfer, functional genes should therefore be inserted into the chromosomal DNA.

The principle function of any inoculation procedure is to perturb the population balance of the rhizosphere against harmful species. However it is crucial that both GEMMOs and wild-types do not reduce the populations of beneficial species. Methodology has not been fully evaluated to make such assessment, but the following serves as to illustrate two of the methods which we have been developing to assist in these evaluations.

4.2 A Most Probable Number (MPN) Recovery Technique

We have isolated *Pseudomonas aureofaciens* as a dominant wild-type bacterium from the leaves and roots of sugar beet plant growing on a clay soil in Oxford (7, 8), and from the leaves and roots of wheat growing on a silt loam in Littlehampton (9). A recombinant *Pseudomonas aureofaciens* strain (SBW25EeZY-6KX) was constructed by site-directed homologous recombination using the *lac*ZY genes encoding lactose utilization (10) and the marker cassette km^r-*xyl*E encoding kanamycin resistance and catechol catabolism (11). The *lac*ZY genes facilitate the cleavage of the chromogenic substrate 5-chloro-4-bromo-3-indolyl-β-D-galacto-pyranoside(X-gal) to yield a blue color. The strategy for the marking of the recipient bacteria was adopted to facilitate detection by simple culture methods which, where necessary, could be enhanced by nucleic acid detection. Further, the introduction of three novel phenotypes ensures the unequivocal identification of the inoculum against a background of indigenous isolates.

In the normal use of the *lac*-ZY marker system, incorporation of kanamycin and X-gal into *Pseudomonas*- selective agar media facilitates the identification of marked colonies carrying the kanamycin resistant and lactose utilisation genes. To evaluate the potential of increasing the sensitivity of plating procedures, an MPN technique was evaluated (12). A culture of the recombinant was counted using conventional plating procedures and a dilution series made. Samples (0.1ml) from each dilution were then pipetted into each of three replicate bottles containing a broth with the *Pseudomonas* – selective medium of Katoh and Itoh (13), with betaine as the selective agent and sole carbon source amended with 50ppm X-gal and 100ppm kanamycin. The impact of soil on the counting procedure was assessed by adding 1% or 10% field soil (w/v) to the dilution series. Bottles were incubated on a rotary shaker at 25 °C for 4 days and assessed for *lac*ZY activity (blue coloration).

Sensitive (one or more recombinant cells added per bottle) and reliable detection was possible in bottles that contained no soil or a 1% (w/v) suspension, where there was a background count from the soil of $>10^6$ non-recombinants per bottle. However sensitivity was lost with the 10% (w/v) soil suspensions. Clearly the technique is very sensitive and simple to use. It can be applied to leaves and roots but then it is better to dislodge the bacteria by sonication, avoiding maceration which would make the medium non-selective by releasing non-selective carbon substrates into the medium. By using 1 litre broth, it is possible to detect as few as one recombinant per 10g of soil.

4.3 The Need for an Eco-Physiological Index (EPI)

Traditional plate counting of individual components in a microbial population is generally too imprecise to measure anything less than a catastrophic change (one or two orders of magnitude). It is therefore a challenge to recover indigenous organisms and assess the impact of GEMMOs. With such a diverse range of phenotypic characters in soil organisms, it is difficult to determine which character to focus on. Winogradsky (14) distinguished bacterial groups as *autochthonous* (where there is a low but steady level of activity on native soil organic matter) and zymogenous (where there is a rapid metabolism of freshly available organic matter). In more modern ecological theory (15), the zymogenous organisms are similar to r-strategists which are basically opportunistic, typically small with rapid growth and high mortality rates. The autochthonous organisms are similar to K-strategists which typically have large size and longevity, slower growth and low mortality rates.

We have attempted to use the simple agar plating method to determine the relative contribution of r- and K-strategists to populations of culturable bacteria (12). Essentially our method depends on the quantification of bacterial colonies as they appear on agar media after one, two, three, four, five, six and ten days. In this way seven counts (or classes) could be generated per plate. We suggest that bacteria exhibiting r-strategies appear early and the K-strategists appear late. In the wheat rhizosphere the population was initially dominated by r-strategists, but as the wheat microbial communities were matured more K-strategists appeared. Soil was dominated by K-strategists. At growth stage 69 (flowering) plants grown in microcosms in the glasshouse were dominated by r-strategists in the rhizosphere whereas the counterparts in the field had a more even balance of r- and K-strategists (Fig 1). The same concepts can be applied to any component member of the bacterial community. For example in our experiments, the *Pseudomonas* populations were affected in a similar manner, except that the differences in distribution patterns between field and glasshouse found on the roots at growth stage 69, were maintained in the soil until the end of the experiment (Fig 1).

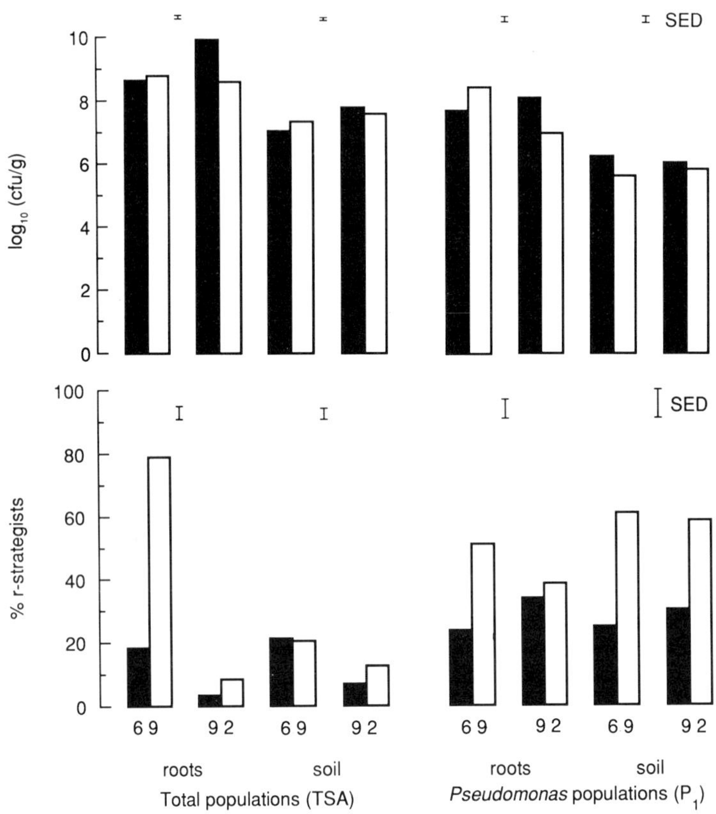

Fig. 1. Total number of bacterial colony forming units (cfu)/g and percentage r-strategists within those populations on wheat roots and in soil. Wheat plants were sampled at flowering (GS69) and ripening (GS92) from the field (solid bars) and from microcosms (open bars) placed in the glasshouse. Tryptone soya agar was used to isolate total populations and P_1 was used as a *Pseudomonas*-selective medium.

The above method of studying changes in r- and K-strategists seems sensitive enough to be able to determine perturbations brought about by impacts on the ecosystem. Previous studies have been based on determinations of impact of single organisms. In this respect, diversity indices such as that proposed by Shannon (16) cannot be applied in our studies *sensu stricto* because groups rather than individual species have been determined. However, the aim is still to determine the eveness of the population and the same equation would seem to apply, but we prefer to call it an "Eco-Physiological Index" (EPI) where:

$$EPI = -\Sigma(p_i \times \log_{10} P_i)$$

with p_i representing each of seven classes as a proportion of the total population in that sample; p_i = population in class i/total population. The more even the distribution of the classes the higher the EPI. It has a maximum value of 0.816 and a minimum of zero for the seven classes under consideration.

The EPI varies with depth and plant growth stage. For example values for the total bacterial populations on roots varied between 0.287 and 0.682 with a standard error of 0.056. With this precision, it gives scope to measure subtle perturbations on the system studied.

4.4 Conclusions

Clearly the crucial factor in determining impact assessment of GEMMOs on rhizosphere population dynamics is the capacity to be able to determine with precision the composition and size of microbial populations. Identification of individuals within a mixed microbial population on general isolation media will always be imprecise unless a very large number, (probably hundreds), of replicate platings or selective media are used. This of course normally only gives counts of culturable genera, such as *Pseudomonas* in our studies. However, the value of such determinations in terms of environmental impact is still questionable. The application of ecological r-K strategy concepts to impact assessment as outlined in this paper is a possible approach to bringing more ecological significance into impact assessment. Another approach would be to determine impacts on functional rather than population biodiversity. Enzyme measurements can usually be made with greater precision than population measurements, but the common problem is that they are difficult to interpret in absolute terms in relation to mineralization or the biocycling of nutrients. These difficulties largely disappear when they are used in a comparative sense, which is required for impact assessment. The crucial information is whether baseline activity is perturbed by the introduction of a GEMMO into the natural microbial community, and whether that had any detremental environmental impact.

Another problem in impact assessment is the collection of large amounts of experimental information which is difficult to interpret. Clearly the help and support of biometricians are important in experimental design and analysis. However there is also a place for the mathematical modeller who can help to ask appropriate questions of experimental systems. A mathematical model has been produced recently (17) which analyses the population dynamics of a GEMMO introduced into the rhizosphere in relation to the soil water potential. The model fits the experimental observations well and there appears to be a great deal of potential in the use of such predictive models in impact assessment, which was the conclusion reached in a recent OECD workshop (18).

Acknowledgements

The work referred to here is part of a co-operative project between Horticulture Research International, NERC Institute of Virology and Environmental Microbiology, King's College London and University of Aberdeen, funded by the Department of the Environment. We acknowledge the help of Emma Sutton, John Fenlon and Ian Thompson on this part of the project.

4.5 References

1. Fry, J.C. and Day, M.J. (eds). Release of Genetically Engineered and other Micro-organisms. 1992 Cambridge University Press, Cambridge UK.
2. Stewart-Tull,, D.E.S. and Sussman, M. (eds) . The Release of Genetically Modified Micro-organisms 1992 Plenum Press, New York.
3. Sun, L., Bazin, M.J. and Lynch, J.M. Plasmid dynamics in a model soil column. Molecular Ecology, *2*, (1993) 9-15.
4. Lilley, A. K., Fry, J. C., Day, M. J. and Bailey, M.J. *In situ* transfer of an exogenously isolated plasmid between *Pseudomonas* spp. in sugar beet rhizosphere. Microbiology, *140* (1994), 27-33.
5. Kerr, A. and Tate, M.E. . Agrocins and the biological control of take-all. Microbiological Science, *1*, (1984) 1-14.
6. Jones, D.A., Ryder, M.H., Clare, B.G., Farrand, S.K. and Kerr, A. . Construction of a Tra^- deletion mutant of pAgK84 to safeguard the biological control of crown gall. Molecular and General Genetics, *212*, (1988) 207-214.
7. Bailey, M.J. and Thompson, I.P. Detection systems for phyllosphere pseudomonads. In: Genetic interactions between Microorganisms in the natural environment. Wellington, E.M.R. & van Elsas, J.D. (Eds.) 1992 pp. 127-141, Pergamon Press.
8. Thompson, I.P., Bailey, M.J., Fenlon, J.S., Fermor, T.R., Lilley, A.K., Lynch, J.M., McCormack, P.J., McQuilken, M., Purdy, K.J., Rainey, P.B. and Whipps, J.M. Quantitative and qualitative seasonal changes in the microbial community from the phyllosphere of sugar beet (*Beta vulgaris*). Plant and Soil, *150* (1993) 177-191.
9. Legard, D.E., McQuilken, M.P., Fenlon, J.S., Fermor, T.R., Whipps, J.M., Thompson, I.P., Bailey M.J. and Lynch, J.M. Evaluation of the microbial phylloplane populations of spring wheat for assessment of the risk to the environment of releasing genetically engineered micro-organisms. In preparation.
10. Drahos, D.J., Henning, B.C. and McPherson, S. Tracking recombinant organisms in the environment: β-galactosidase as a selectable non-antibiotic marker for fluorescent pseudomonads. Bio/Technology, *4*, (1986) 439-444.
11. Williams, P.A. and Murray, K. Metabolism of benzoate and the methyl benzoates by *Pseudomonas putida* (Arvilia) mt-2: evidence for the existence of a TOL plasmid. Journal of Bacteriology, *125*, (1974) 416-423.
12. Leij, F.A.A.M, de, Bailey, M.J., Whipps, J.M. and Lynch, J.M. A single most probable number tech-

nique for the sensitive recovery of a genetically-modified *Pseudomonas aureofaciens* from soil. Letters in Applied Microbiology, *16*, (1993) 307–310.

13. Katoh, K. and Itoh, K. New selective media for *Pseudomonas* strains producing fluorescent pigment. Soil Science and Plant Nutrition, *29*, (1983) 525–532.
14. Winogradsky, S. Sur la microflore autochtone de la terre arable. Comptes rendues ebdomadrire des seances de l'Academie des Sciences (Paris) D, *178*, (1924) 1236–1239.
15. Southwood, T.R.E. Bionic strategies and population parameters. In: Theoretical Ecology, Principles and Applications. (ed. R.M. May), 1976 pp 26–48, Blackwell, Oxford.
16. Shannon, C.E. A mathematical theory of communication. Bell Systems Technology, *27*, (1948) 379–423.
17. Scott, E.M., Rattray, E.A.S., Prosser, J.I., Killham, K., Glover, L.A., Lynch, J.M. and Bazin, M.J. A mathematical model for dispersal of bacterial inoculants colonising the wheat rhizosphere. Soil Biology and Biochemistry, in Press.
18. Bazin, M.J. and Lynch, J.M., (eds) Terrestrial Gene Exchange: Mathematical Modelling and Risk Assessment, Chapman and Hall, London, in Press.

5 Developing Concepts in Biological Control: A Molecular Ecology Approach

David N. Dowling, Bert Boesten, Paul R. Gill Jr. and Fergal O'Gara

5.1 Introduction

The application of molecular biology techniques and tools to the study of microbial ecology has led to renewed interest in this field. An understanding of the ecology of a microorganism is a fundamental requirement for the successful introduction of a microbial inoculant into the open environment. This is particularly true for the biological control of root pathogens in the rhizosphere, where one is actively seeking to alter the ecological balance so as to favour growth of the host plant and to curtail the development of pathogens.

In the past ten years considerable progress has been achieved using molecular genetic techniques to elucidate the important microbial factors or genetic traits involved in the suppression of fungal root diseases (1–8). In our laboratory, we have focussed on the role of fluorescent siderophores in the ecological competence of inoculant strains and the role of antifungal metabolites in the suppression of *Pythium ultimum* induced "damping-off" of sugarbeet seedlings. Our approach (see fig 1), in common with other workers (see Chapters 1 and 6, this volume), has been to isolate mutants defective in biocontrol and their corresponding wildtype sequences.

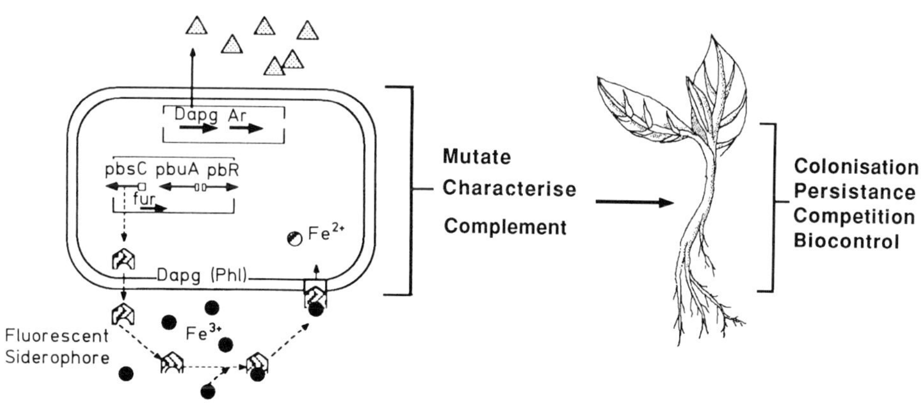

Fig. 1. Outline strategy to elucidate key genetic traits of fluorescent pseudomonads implicated in rhizosphere competence and biological control.

Once complementing clones are available and the loci characterised, it is possible to evaluate the role these gene(s) play in the biocontrol ability and rhizosphere competence of the inoculant. Biological control is multifactoral in nature and it has been shown that mixed strains are often superior to single inoculants (9). This is presumably due to the greater genetic diversity of microbial consortia, leading to a potential increase in available disease suppressive mechanisms. When key biocontrol traits are identified, the biotechnologist can make a rational choice in the selection of strains with specific traits for a particular field application. Potentially useful genetic traits such as; antifungal metabolite synthesis may be introduced into heterologous strains to increase the genetic potential of single biocontrol agents without decreasing the fitness or ability to compete in the rhizosphere. Introduction of genes into strains for environmental release requires the use of specialised vector systems (see chapter 7, this volume) and in consequence part of our research programme has been directed towards the development of biologically contained, and environmentally friendly plasmid based vector system.

5.2 Siderophore-Mediated Competitive Exclusion of Phytopathogens

Iron is an essential nutrient for most microorganisms, however it is probably unique in its near quantitative insolubility in aerobic environments at pH 7 (10). This imposes a severe limitation on its availability. To overcome this, most microorganisms secrete low molecular weight high-affinity iron chelating compounds termed siderophores. Although the functional groups within microbial siderophores that are involved with the chelation of iron are more conserved, the overall structures of these compounds varies widely. *Pseudomonas* strains usually produce a siderophore consisting of a fluorescent quinoline group covalently linked to a peptide in which two of the amino acid side chains are modified for iron chelation. There is typically a strain dependent variation in the structure of the siderophore produced. This family of structurally related siderophores are generally referred to as pseudobactins or pyoverdine-type siderophores. In the prototypical pseudobactin type siderophore, the iron-binding groups consist of the two hydroxyl groups of the quinoline moiety, reminiscent of a catechol, an alpha-hydroxyl carboxylic acid (a modification of the side chain of aspartate) and a hydroxymate moiety (as a modification of ornitine) (8).

The second component of the standard microbial iron transport system is a receptor that specifically recognizes the ferric complex of the siderophore for uptake of the metal. In Gram-negative bacteria, such receptor proteins are found in the outer membrane (11). A substantial amount of research in biological control has focused on the potential of using or modifying bacterial iron transport systems to enhance the abilities of Plant Growth Promoting Rhizobacteria (PGPR) strains. It has been found that a given *Pseudomonas* strain will typically be able to recognize not only its own pseudobactin siderophore, but in varying degrees, also those produced by other pseudomonad strains (10,11,12). Specific uptake of the ferric-siderophore complex by a receptor is thought to be dependent on the composition and sequence of the amino acids comprising the peptide component and modifications of the quinoline moiety of the siderophore. Selective enrichment of neutral and PGPR strains in the rhizosphere based on iron availability mediated by the appropriate siderophore receptor, may be a primary mechanism by which the growth of deleterious and pathogenic microbes is attenuated. One experimental objective in our research programme is to enhance the ability of recognised biocontrol strains to aquire iron solubilized by a variety of pseudobactin siderophores in a soil, by providing them with additional pseudobactin-type receptors. It will be interesting to evaluate whether such engineered biocontrol strains could have improved persistence and competitive performance in the rhizosphere.

Pseudomonas sp. strain B24, a rhizosphere isolate, is unable to use many different pseudobactin-type siderophores produced by other *Pseudomonas* strains isolated from various Irish soils. To evaluate the possibility that an additional pseudobactin receptor might increase the potential of this strain to colonize roots and persist in the rhizosphere, the primary receptor gene from *Pseudomonas* sp. strain M114 (13) was introduced into strain B24 on a plasmid vector. Initially, it was found that the ability of the engineered strain to use additional pseudobactin siderophores was significantly enhanced (14). Of approximately 250 fluorescent rhizophere *Pseudomonas* strains tested, 32% were able to crossfeed strain B24 under iron-limiting conditions. In this experiment, it was shown that strain B24 had the ability to use as an iron source, the pseudobactin siderophore produced by a limited number of strains. In contrast strain B24 harboring the strain M114 siderophore receptor gene could be crossfed by 82% of the same test strains. Thus, introduction of the M114 receptor gene into strain B24 increased its ability to use siderophores produced by an additional 50% of the test strains used in this study. In a series of in vitro studies it was found that the engineered B24 strain had increased competitive ability compared to the parental B24 strain under low iron concentrations. This effect was not evident when iron was not limiting.

A question remains as to the identity of the test strains and the variation, if any, of the structures of the pseudobactins that they produce. Preliminary evidence, using an arbitrarily primed-polymerase chain reaction (AP-PCR) DNA fingerprinting method (15), suggests that a collection of rhizosphere strains, that are able to crossfeed the engineered B24 strain, produce different fingerprint patterns. This indicates that these are genotypically diverse and may produce different siderophores.

The expression of iron-regulated genes, such as siderophore biosynthetic and siderophore receptor genes, is modulated by the level of available environmental iron. In *Pseudomonas* sp. strain M114, iron-regulated gene expression is modulated by both positive and negative regulatory elements (16,17). A positive regulatory locus has been identified by insertion mutagenesis in strain M114, and it contains an open reading frame that encodes a protein sequence with homology to known DNA-binding proteins (18). Under conditions of iron limitation, this gene *(pbrA)* is required for iron-regulated gene expression. Iron regulated gene expression in *Escherichia coli*, on the other hand is controlled by a single negative regulatory element that acts as a classical repressor and functions together with ferrous iron as its co-repressor.

Transcription of iron regulated genes is blocked under conditions of iron sufficiency. In *E. coli* a consensus DNA sequence 5′ to the coding sequences of iron regulated genes has been shown to be the site at which the repressor binds to inhibit transcription (19). Because similar "iron box" sequences are found 5′ to *pbrA* from strain M114 it may indicate that *pbrA*, in turn, is regulated by a Fur-like protein in strain M114. There is evidence that *Pseudomonas* sp. strain M114 contains a fur-like gene and such a gene has recently been isolated and sequenced from *P. aeruginosa* (20).

Having identified a positive regulatory element that is sufficient for expression of an iron-regulated M114 promoter in a heterologous background (16), we may be in a position to engineer diverse rhizosphere strains with improved iron assimilation ability. This may lead to inoculants with improved biocontrol and competitive properties.

5.3 Exploiting Antifungal Metabolites to Enhance Biological Control

Another mechanism by which a biocontrol agent acts to protect plants from pathogens depends on the production of antifungal metabolites (4,6,7). Growth inhibition of soil borne fungal phytopathogens (unrelated to siderophore-mediated inhibition) by fluorescent rhizosphere pseudomonads has been demonstrated in vitro (Petri dish bioassays) in a number of studies (7). In a limited number of cases, production of antibiotic-like compounds by a given *Pseudomonas* strain is strongly correlated with the ability of that strain to control a fungal phytopathogen in field tests or gnotobiotic assays (6). The production of the antimicrobial compound, 2,4-diacetylphloroglucinol (DAPG), by *Pseudomonas fluorescens* CHA0 was previously demonstrated to be an important factor in the control of black root rot on tobacco (21). *Pseudomonas* sp. strain F113, studied in our laboratory, also produces DAPG, and acts as a biocontrol agent for sugarbeet by reducing the incidence of *Pythium* induced "damping-off" in seed germination tests. A mutant of strain F113 deficient in the production of DAPG, i.e., strain G22, was unable to protect this plant from the phytopathogen (22). A chromosomal region isolated from strain F113 was capable of complementing the mutation in strain G22, in that its ability to produce DAPG was restored. A role for the biosynthesis of DAPG by this complementing region was suggested because cell extracts of strain G22 were unable to convert 2-monoacetylphloroglucinol (MAPG) into DAPG whereas strain F113 and the complemented mutant were able to catalyze this conversion. From these studies a biosynthetic pathway for DAPG production by strain F113 was proposed (23).

A desirable goal is to construct improved biocontrol agents by means of incorporating additional factors that antagonize the growth of a target pathogen into a single strain. Therefore, the chromosomal region from F113 encoding the conversion of MAPG to DAPG was introduced into a series of rhizosphere *Pseudomonas* isolates to determine if such isolates were now capable of DAPG production. Of 11 strains tested, only strain M114 was capable of DAPG production after introduction of the F113 chromosomal region (22). A second rhizosphere *Pseudomonas* isolate, E1/7, previously incapable of producing DAPG is capable of doing so when provided with both the region required for the conversion of MAPG to DAPG and an additional regulatory region from strain F113 (A. Fenton, unpublished data).

It should be noted that DAPG can be phytotoxic at high concentrations (24) and strains that have been engineered to overproduce DAPG show a phytotoxic effect on susceptible plant species (25). Therefore regulated gene expression is likely to be an essential feature for applications involving the microbial production of DAPG.

5.4 Stability of Introduced Genes and Biological Containment Systems for GMO's

When genes, or gene clusters are to be introduced into a microorganism (be it by intergration into the chromosome or by introduction of self-replicating plasmids), a positive selectable phenotype is essential to distinguish the modified microorganism from the parent population. In the case of genetic instability, a positive selective pressure may have to be maintained throughout the life cycle of the GMO to maintain the engineered trait. Occasionally the introduced gene itself will provide a selectable phenotype. For instance in the case of metabolic genes, it may allow the GMO to utilize specific compounds as a carbon source which cannot be metabolised by the parent organism. Minimal medium supplemented with that compound as a sole carbon source, will only allow the GMO to grow. However, in a complex environment such as the soil or the rhizosphere, alternative carbon sources will be available to sustain growth of the parent microorganism and the GMO will have no selective advantage. The burden of maintaining the extra genetic material may even result in a disadvantage for the GMO in such highly competitive surroundings. Similarly, a gene that codes for a mechanism that renders the GMO resistant to toxic compounds (such as antibiotics or heavy metals), will only provide a selective advantage to the GMO when such compounds are present in the environment.

Self-replicating plasmids provide a convenient and efficient vector system. It is simple to introduce a gene on a plasmid, than to stably integrate it into the genome, without interrupting (or influencing) vital functions of the new host. In the case of self-replicating plasmids stability becomes a major consideration. However, in some cases loss of plasmids and therefore the introduced genes does not have to be a disadvantage to the overall performance of a microbial inoculant One can imagine a scenario where loss of introduced genes after an initial period, such as; during germination and early development of seedlings would be an advantage. Such an inoculant would loose its "GMO genotype" and revert to its "wild type" status. However, when plasmids are used to introduce "marker" genes for monitoring purposes, 100% plasmid stability is essential.

The key to plasmid maintenance, is the ability to apply a positive selective pressure. Under laboratory conditions, antibiotics in conjunction with resistance genes are commonly used for this purpose. In the field this type of selective pressure cannot be easily sustained. The wisdom of introducing antibiotic resistance genes into the environment is also under debate. A solution to this problem is to combine a host microorganism which is mutated in an essential gene with a plasmid, bearing a copy of that particular gene. This leads to a host/vector combination that is intrinsically stable under normal growth conditions.

This approach was utilized in our laboratory for the construction of an antibiotic resistance free vector system (26). The plasmid vector (pGDT11) contains a copy of the *Lactococcus lactis* thymidylate synthase (*thyA*) gene. Thymidylate synthase (TS) is a key enzyme in de novo DNA synthesis and essential for growth of the organism in the absence of an external supply of thymidine (or thymine). These compounds are rare in natural environments and the survival of microorganisms mutated in this gene is greatly impaired. Plasmids containing a copy of the *thyA* gene can be readily selected in this background, because only bacteria containing such a plasmid are viable. In our hands, the combination of a *Rhizobium meliloti* thy$^-$ host and pGDT11 ensures absolute plasmid stability under laboratory, greenhouse and field conditions. Therefore pGDT11 and similar plasmids provide a suitable vector system to introduce genes of interest into inoculant strains and ensure stability of the GMO throughout the life cycle of the strain.

The limited ability for survival of microorganisms mutated in thymidylate synthase provides an opportunity to develop biological containment systems for microbial inoculants in the open environment. The concept of biological containment implies the ability to limit the spread and persistence of an inoculant in the field. Ideally an inoculant should only persist long enough to perform whatever task it is intended for (eg. protect the developing root system of seedlings against attack by fungal pathogens). After that, their numbers should decline to undetectable levels. Biologically contained inoculants will avoid problems such as have been encountered in the past with highly persistent and competitive inoculants which can prevent the subsequent use of more efficient strains. In fact, the use of contained inoculants would require fresh inoculations each growing season. This would allow the grower to use the latest and most efficient inoculants available. The timespan of exposure of a GMO to the environment would be limited which could reduce the chance of exchange of genetic material between the GMO and resident microorganisms. Finally, the population decline of an inoculant introduced in high numbers in the environment, could facilitate the restoration of any alterations to the resident microbial populations.

Currently, we are developing a biological containment system for *Rhizobium* inoculants based on the limited survival of *R.meliloti* Thy mutant strains. There are three aspects to this work: The first requirement is for stable *R.meliloti* Thy mutants which are unable to revert back to a wildtype phenotype. This can be achieved by a "reverse genetics" approach. The *R.meliloti* Thy gene has been cloned and mutated by deletion of an internal piece of DNA. This mutated gene can be then "marker exchanged" into the genome of a number of *Rhizobium* strains. Secondly, the survival of thy mutants in various conditions from rooting solution in the laboratory, to soil in the greenhouse and in the presence and absence of host (and non-host) plants were monitored. Such studies indicated that these mutants do not survive in the open environment but are capable of colonising the root system of the alfalfa host plant. This would be an ideal situation because an inoculant would only survive in association with the host plant. Unfortunately, *R.meliloti* thy mutants do not nodulate. Therefore the third strand of this scheme is to develop conditional expression systems of the thy gene. The aim is to express the thy gene from a regulated promoter which is only active under those conditions where viability of the inoculant is required. In the case of a *Rhizobium* inoculant, for example it will be necessary to express the gene during nodulation and symbiosis. Promoters from nodulation and symbiotic genes are currently under investigation for this purpose.

5.5 Conclusion

Biological control of rhizosphere phytopathogens and deleterious microbes, using selected *Pseudomonas* isolates, has been difficult to implement on a reproducible basis. The reasons for this are no doubt complex, variable and includes both biotic and abiotic factors. The method used for the introduction of a biocontrol agent could also play a crucial role in their ability to perform. These methods must be also compatible with established farming practices. Strategies utilizing molecular genetic techniques have been developed to complement ongoing research ranging from the characterization and genetic improvement of a selected biocontrol agent to the measurement of its persistance and dispersal. Methods used for the characterization and genetic improvement of a strain may be used to define the mechanism by which pathogens and deleterious microbes are controlled and a detailed understanding of these mechanisms, by which a biocontrol agent functions, will allow the development of more refined methods for agricultural applications. Finally, biocontrol should be considered as part of a disease control strategy like Integrated pest management, (IPM) (27) which offers a successful approach for the deployment of both agrochemicals and biocontrol agents.

Acknowledgements

We would like to thank the members of the plant/micobial interactions group at UCC for the inclusion of unpublished data and Mr P.Higgins for technical assistance. This work was supported in part by contracts from the European Community [ECLAIR AGRE 0019-C; BRIDGE BIOT-CT90-0166-C, BIOT-CT91-0293, BIOT-CT91-0283; FLAIR AGRE-0019-C; BIOTECH BIO2-CT93-0053, BIO2-CT93-0196, BIO2-CT92-0084]

5.6 References

1. Weller, D.M. Biological control of soilborne plant pathogens in the rhizosphere with bacteria. Annu. Rev. Phytopathol. 26 (1988) 379–407.
2. Lockwood, J.L. Evolution of concepts associated with soilborne plant pathogens. Annu. Rev. Phytopathol. 26 (1988) 93–121.
3. Bull, C.T. and Weller, D.M. and L.S. Thomashow. Relationship between root colonization and suppression of *Gaeumannomyces graminis* var. tritici by *Pseudomonas fluorescens* strain 2–79. Phytopathology 81 (1991) 954–959.
4. O'Sullivan, D.J., and F. O'Gara. Traits of fluorescent *Pseudomonas* spp. involved in suppression of plant root pathogens. Microbiol. Rev. 56 (1992) 662–676.
5. Schippers, B. A.W. Bakker and P.A.H.M. Bakker. Interactions of deleterious and beneficial rhizosphere microorganisms and the effect of cropping practices. Annul. Rev. Phytopathol. 25 (1987) 339–358.
6. Gutterson, N.I. Microbial fungicides: recent approaches to elucidating mechanisms. Critical Rev. in Biotechnology. 10 (1990) 69–91.
7. Fravel, D. Role of antibiotics in the biocontrol of plant diseases. Ann. Rev. Phytopathol. 26 (1988) 75-91.
8. Leong, J. Siderophores: their biochemistry and possible role in the biocontrol of plant pathogens. Ann. Rev. Phytopathol. 24 (1986) 187–209.
9. Pierson, E.A., and D.M. Weller. Recent work on control of take-all of wheat by by fluorescent pseudomonads. In: C. Keel, B. Koller and G. Defago (Eds.), Plant growth-promoting rhizobacteria-progress and prospects, WPRS Bulletin 1991/XIV/8, pp. 96–97.
10. Neilands, J.B. Siderophores of bacteria and fungi. Microbiol. Sci. 1 (1984) 9–14.
11. Neilands, J.B. Microbial envelope proteins related to iron. Ann. Rev. Microbiol. 36 (1982) 285–309.
12. Buyer, J.S. and J. Leong. Iron transport-mediated antagonism between plant growth-promoting and plant-deleterious *Pseudomonas* J. Biol. Chem. 261 (1986) 791–794.
13. Morris, J., D.J. O'Sullivan, M. Koster, J. Leong, P. Weisbeek and F. O'Gara. Characterization of fluorescent ferric siderophore-mediated iron uptake in *Pseudomonas* sp. strain M114: evidence for the existence of an additional ferric siderophore receptor. Appl. Environ. Microbiol. 58 (1992) 630-635.
14. Dowling, D.N., B. Boesten, D.J. O'Sullivan, P. Stevens, J. Morris and F. O'Gara. Genetically modified plant-microbe interacting strains for potential release into the rhizosphere, pp. 408–414. In: E. Galli, S. Silver and B. Witholt (eds.), Pseudomonas Molecular Biology and Biotechnology. American Society for Microbiology, Washington, D.C. (1992).
15. Welsh, J. and M. McClelland. Genomic fingerprinting using arbitrarily primed PCR and a matrix of pairwise combinations of primers. Nucleic Acids Research 19 (1991) 5275–5279.
16. O'Sullivan, D.J., and F. O'Gara. Regulation of iron assimilation: nucleotide sequence analysis of an iron-regulated promoter from a fluorescent pseudomonad. Mol. Gen. Genet. 228 (1991) 1–8.
17. O'Sullivan, D.J. and F. O'Gara. Iron regulation of ferric iron uptake in fluorescent pseudomonads: cloning of a regulatory gene. Mol. Plant-Microbe Interact. 3 (1990) 86–93.
18. Sexton,R., Gill,P.R.,Jr., Callanan,M.J, O,Sullivan,D.J., Dowling,D.N. and F.O'Gara. Transcriptional activation is required for iron responsive expression in *Pseudomonas* sp.Strain M114: Cloning and characterisation of the transcriptional activator pbrA and heterologous expression in *E.coli.* (Submitted for publication)
19. deLorenzo, V. S. Wee, M. Herrero and J.B. Neilands. Operator sequences of the aerobactin operon of plasmid Col;V-K30 binding the ferric uptake regulation (fur) repressor. J. Bacteriol. 169 (1987) 2624–2630.
20. Prince,R.W., Cox,C.D. and M.L.Vasil. Coordinate regulation of siderophore and exotoxin A production: molecular cloning and sequencing of the *Pseudomonas aeruginosa fur* gene J. Bacteriol. 175 (1993) 2589–2598.
21. Keel, C., P. Wirthner, T. Oberhänsli, C. Voisard, U. Burger, D. Haas, and G. Défago. Pseudomonads as antagonists of plant pathogens in the rhizosphere: role of the antibiotic 2,4-diacetylphloroglucinol in the suppression of black root rot of tobacco. Symbiosis *9* (1990) 327–341.

22. Fenton, A.M., P.M. Stevens, J. Crowley, M. O'Callaghan and F. O'Gara. Exploitation of gene(s) involved in 2,4-diacetylphloroglucinol biosynthesis to confer a new biocontrol capability to a *Pseudomonas* strain. Appl. Environ. Microbiol. 58 (1992) 3873-3878.
23. Shanahan, P., J.D. Glennon, J.J. Crowley, D.F. Donnelly and F. O'Gara. HPLC assay of microbially derived phloroglucinol antibiotics in establishing the biosynthetic route to production, and the factors affecting their regulation. Analytica Chimica Acta. 272 (1993) 271-277.
24. Shanahan, P. Analytical biotechnology of antimycotic metabolites from fluorescent pseudomonads. Ph.D. dissertation, University College Cork, Cork, Ireland. (1992).
25. Maurhofer, M., C. Keel, U. Schnider, C. Voisard, D. Haas, and G. Défago. Influence of enhanced antibiotic production in *Pseudomonas fluorescens* strain CHA0 on its disease suppressive capacity. Phytopathology *82* (1992) 190-195.
26. Ross, P., F. O'Gara and S. Condon. Thymidylate synthase gene from Lactococcus *lactis* as a genetic marker: an alternative to antibiotic resistance genes. Appl. Environ. Microbiol. 56: (1990) 2164-2169.
27. Flint, M.L. and R. van den Bosch. Introduction to Integrated Pest Management. Plenum Press, New York. (1981)

6 Biocontrol of Root Diseases by *Pseudomonas fluorescens* CHA0: Current Concepts and Experimental Approaches

Christophe Voisard, Carolee T. Bull, Christoph Keel, Jacques Laville, Monika Maurhofer, Ursula Schnider, Geneviève Défago and Dieter Haas

6.1 Introduction

Certain strains of fluorescent pseudomonads can effectively colonize plant roots and protect plants from diseases caused by a variety of root pathogens. Such beneficial or plant health-promoting strains are emerging as promising biocontrol agents. They are suited as soil inoculants either individually or in combination and may be compatible with current chemical pesticides (1, 2, 3, 4, 5, 6, 7, 8). In our biocontrol studies, we have focused on *Pseudomonas fluorescens* strain CHA0, an isolate from a suppressive soil in the western part of Switzerland (9). This strain was originally shown to colonize tobacco roots and to suppress black root rot, which is caused by the fungus *Thielaviopsis basicola* (9, 10). Subsequent work has established that disease suppression by strain CHA0 displays little specificity with respect to the host plant and the pathogen. Protected plants include wheat, cucumber, sugar beet, cotton, flax, corn, and cress. Pathogenic action of at least the following fungal pathogens can be reduced by strain CHA0: *Pythium ultimum, Gaeumannomyces graminis* var. *tritici* (Ggt), *Fusarium oxysporum* f.sp. *cucurbitaceae*, *Phomopsis sclerotioides*, and *Rhizoctonia solani* (11, 12, 13, 14, 15; our unpublished data). Since the interactions between *P. fluorescens*, other organisms and the soil environment are extremely complex, it became important to develop reproducible methods that allow us to monitor the plant-beneficial effects of strain CHA0 reliably and to analyze the traits that make it an effective biocontrol agent. In section 6.2 we will review some of our approaches to investigate the mechanisms by which strain CHA0 achieves biological control.

We are using strain CHA0 as a model organism to study not only the mechanisms of disease suppression but also the ecological impact of introduced plant-beneficial bacteria (see section 6.3). In a parallel approach, we are investigating the potential applications of biological control agents to improve the yields of protected crops. We are testing a variety of strains, singly and in combination, for the development of greenhouse applications. A brief account of this work is presented in section 6.4.

6.2 Mechanistic Studies on Biocontrol Traits of *Pseudomonas Fluorescens* CHA0

6.2.1 Chemical Identification of Extracellular Metabolites

Metabolites produced and excreted by *P. fluorescens* are assumed to be important biotic factors in the biological control of root diseases (2, 5, 16, 17, 18, 19, 20, 21). Until now, about a dozen low molecular weight compounds have been identified in culture supernatants of *P. fluorescens* CHA0 (Table 1). These products can be broadly classified into two groups: siderophores and secondary metabolites.

The siderophores (e.g., iron chelators) pyoverdine (pseudobactin), salicylate and pyochelin are all produced by *P. fluorescens* CHA0 (22, 23; our unpublished results) and by other fluorescent pseudomonads when these bacteria are grown under iron-limiting conditions (24; reviewed by Loper & Buyer [20] and O'Sullivan & O'Gara [5]). The fluorescent siderophore pyoverdine of strain CHA0 has been isolated and shows spectral properties that are similar to those of pyoverdines from other fluorescent pseudomonads (22). The chemical composition of CHA0 pyoverdine has not been determined but this compound presumably differs in its peptide moiety from the chemically defined pyoverdine of *P. aeruginosa* PAO1 (23). Salicylate, a precursor of pyochelin (25), was shown to be a product of *P. fluorescens* CHA0, by comparison with the pure chemical (23). Pyochelin was tentatively identified based on its co-elution with authentic pyochelin of *P. aeruginosa* during HPLC analysis (our unpublished results).

The production of three siderophores by strain CHA0 and other fluorescent pseudomonads might give these bacteria a competitive advantage over other rhizosphere microorganisms, especially pathogens, in iron acquisition. The amount of Fe^{3+} that is available to microorganisms in well-aerated soils is restricted by the low solubility of Fe^{III} oxide-hydroxide and, according to the siderophore hypothesis (20, 21), the microorganisms possessing the most efficient Fe^{3+} uptake systems should have a growth advantage and out-compete deleterious microorganisms. We would like to point out here that two siderophores produced by strain CHA0 may have further functions. Salicylate, in addition to being a siderophore (23, 26), is a plant hormone and inducer of systemic resistance against pathogen attack (27, 28, 29). Pyochelin, the third iron chelator of strain CHA0, also complexes several divalent ions (Table 1) and may promote their uptake in fluorescent pseudomonads (24, 30).

When growth slows down and cells enter stationary phase, *P. fluorescens* CHA0 produces a battery of secondary metabolites (Table 1). Hydrogen cyanide (HCN) synthesized by strain CHA0 was identified by three independent chemical methods (22, 33). The other metabolites – indoleacetate (IAA), 2,4-diacetylphloroglucinol (Phl), pyoluteorin (Plt) and pyrrolnitrin (Table 1) – were isolated from growth media by solvent extraction after acidification, purified by HPLC and identified chemically. The biological properties of these secondary metabolites are diverse. Cyanide

Tab. 1. Metal Chelators and Secondary Metabolites Excreted by *P. fluorescens* CHA0

Extracellular metabolite	Characteristics	Role in biological control under gnotobiotic conditions	Reference
Pyoverdine (Pseudobactin)	Fluorescent siderophore (Fe^{3+}), chelator of Al^{3+}, Cu^{2+}, Mn^{2+}, etc.	No definitive role has been assigned; a pyoverdine-negative mutant has wild-type capacity to suppress black root rot of tobacco and take-all of wheat	10, 22, 31
Salicylate	Siderophore, inducer of systemic resistance in plants	ND[a]	23
Pyochelin	Siderophore, chelator of Cu^{2+}, Co^{2+}, Ni^{2+}, etc.	ND	24; our unpublished results
Hydrogen cyanide	Biocide, chelator of Fe^{2+}, Cu^{2+}, Co^{2+}, Ni^{2+}, etc.	An *hcn* deletion mutant has reduced capacity to protect tobacco from black root rot, but protects wheat from root diseases normally	32, 33
Indoleacetate (IAA)	Plant growth hormone	TSO-negative mutants produce less IAA at pH <7 but suppress black root rot of tobacco and take-all wheat normally	34
2,4-Diacetyl-phloroglucinol[b]	Antibiotic, phytotoxic compound	A *phl*::Tn*5* mutant has reduced capacity to suppress black root rot of tobacco and take-all of wheat	12, 13
Pyoluteorin	Antibiotic, phytotoxic compound	A *plt*::Tn*5* mutant has reduced capacity to suppress *Pythium*-induced disease of cress	15; our unpublished results
Pyrrolnitrin	Antibiotic	ND	Our unpublished results
X[c]	"Autoinducer"	ND	Paul Williams and Gordon Stewart, personal communication

[a] ND, not determined
[b] Phloroglucinol and monoacetylphloroglucinol have also been detected in culture supernatants.
[c] HPLC fractions of culture supernatants are capable of activating the *luxRAB* reporter system in *E. coli* (35); X appears to be similar but not identical to KHL.

is toxic in general because it forms stable complexes with several divalent metal ions (Table 1); in particular, cytochrome c oxidase of many organisms is strongly inhibited by cyanide (36).

Fluorescent pseudomonads and a variety of other microorganisms, however, are relatively insensitive to cyanide; they have an alternate cyanide-resistant cytochrome oxidase. Cyanide-resistant respiration develops when cells divide slowly or cease to grow (37, 38, 39). The concentrations of HCN (~ 100 μM) that are produced *in vitro* by *P. fluorescens* CHA0 at the end of the exponential growth phase are insufficient to cause growth inhibition of the producer strain and of other fluorescent pseudomonads but inhibit growth of some fungi, e.g. *Thielaviopsis basicola* and *Septoria tritici* (22, 33; M. Flaishman, personal communication).

IAA, a plant growth hormone (Table 1), is derived from L-tryptophan via at least two pathways in *P. fluorescens* CHA0. The pathway initiated by tryptophan side chain oxidase (TSO) is induced in stationary phase (34). The quantities of IAA that strain CHA0 excretes in liquid cultures are lower than those observed for some other rhizosphere microorganisms, e.g. *Azospirillum* spp. (34, 40, 41).

Three antibiotics have been identified as stationary-phase products of *P. fluorescens* CHA0: Phl, Plt and pyrrolnitrin (Table 1). They all have broad-spectrum antibacterial and antifungal activities (12, 13, 42, 43, 44, 45, 46, 47). Antibiosis has been proposed to be a critical factor in biocontrol (1, 2, 5, 16, 18). It is striking that at least two of the antibiotics, Phl and Plt, are also phytotoxic at concentrations that are inhibitory to fungi and some bacteria, excluding the producer strains (13, 15, 48).

The recent discovery of a compound (X) similar to N-(ß-ketocaproyl)-L-homoserine lactone (KHL; Table 1) in culture supernatants of *P. fluorescens* CHA0 (P. Williams and G. Stewart, personal communication) is particularly intriguing. KHL is better known as the autoinducer of bioluminescence in *Vibrio fischeri*. When this marine bacterium reaches a certain cell-density in batch culture, autoinducer accumulates in the medium. Above a threshold concentration (~1 ng/ml), KHL triggers transcription of the *lux* genes encoding the enzymes involved in bioluminescence. Besides autoinducer, environmental signals (e.g., low O_2 and Fe levels) contribute to the expression of bioluminescence (49, 50, 51, 52). Recently, KHL has also been detected and implicated as an important signal of secondary metabolism and virulence factors in *Erwinia carotovora, Serratia marcescens, Pseudomonas aeruginosa* and several other bacterial species (35, 53). KHL is structurally similar to an extracellular autoregulator of *Streptomyces* species, termed A-factor, which is an essential signal for antibiotic synthesis and cellular differentiation (54). A fascinating picture is emerging: Many microorganisms, when they reach high cell densities, produce small quantities of chemical signals which trigger very specifically a range of cellular functions: differentiation, antibiosis, symbiosis, or virulence (52, 54).

6.2.2 Genetic Manipulation of Strain CHA0

Which of the extracellular siderophores and secondary metabolites of *P. fluorescens* might be involved in plant protection? To examine this question we have begun to isolate mutants affected in the production of some of these metabolites. One advantage of this approach is that we can test the biocontrol abilities of strains having a defined phenotype, i.e., lack or overproduction of a particular extracellular metabolite. Success of this approach depends on our ability to isolate and identify metabolites involved in biocontrol. Another experimental strategy would be to screen a collection of *P. fluorescens* mutants for reduced suppressive ability. This approach could be hampered by the difficulty to identify the biochemical defects causing loss of suppression and has not been attempted by our laboratory. For a more thorough discussion of the strategies used to investigate biocontrol mechanisms of soilborne pathogens the reader is referred to the review by Handelsman & Parke (18).

Mutants of *P. fluorescens* CHA0 affected in secondary metabolism are formed spontaneously (see section 6.2.6) or can be isolated after transposon insertion mutagenesis. We found that an effective delivery system for Tn5 insertion in strain CHA0 is provided by the ColI1 plasmid pLG221, which is derepressed for conjugative transfer and unable to replicate in *Pseudomonas* species (55). An alternative delivery system is based on the IncP plasmid pME305, a deletion derivative of RP1 (56). For unknown reasons, wild-type RP1 is transmissible at very low (or undetectable) frequencies in crosses with *P. fluorescens* CHA0 recipients whereas pME305 overcomes this conjugational barrier (56). Since pME305 is also temperature-sensitive for replication, it can be used as a delivery plasmid for transposons. To achieve elimination of pME305, several rounds of colony purification may be required at 35 °C (the maximal temperature for growth of strain CHA0) because some replication of pME305 is still possible at this temperature. Several transposons have been loaded onto pME305 (56, 57). Of these, Tn*5-259* and Tn*1733* have been used for mutagenesis of strain CHA0 (Table 2).

Tab. 2. Plasmids used for genetic manipulation of *P. fluorescens* CHA0

Plasmid	Characteristics	Use	Reference
pVK100	IncP Km Tc Mob Tra$^-$, 23 kb	Cloning vector; cosmid suitable for construction of genomic libraries	33, 65
pME305	IncP Ap Tc Rep(ts) Tra, 48 kb	Mobilizing plasmid, suitable for use in strain CHA0	56
pME497	IncP Ap Rep(ts) Tra, 47 kb	Mobilizing plasmid, suitable for mobilization of pVK100 and its derivatives	56
pLG221 = ColIb*drd-1 cib*::Tn*5*	IncI 1 Km Tra, 99 kb	Suicide plasmid for Tn*5* mutagenesis in strain CHA0	13, 33, 55
pME12 = pME305::Tn*5-259*	IncP Ap Tc Km Hg Rep(ts) Tra, 56 kb	Delivery plasmid for Tn*5-259* mutagenesis in strain CHA0	33, 56
pRU672 = pME305::Tn*1733*	IncP Ap Tc Km Rep(ts) Tra, 57 kb	Delivery plasmid for Tn*1733* mutagenesis in strain CHA0	10, 57
pME3087, pME3088	ColE1 Tc Mob, 6.9 kb	Mobilizable suicide plasmids for gene replacements in strain CHA0	61; Fig. 1
pME3049	ColE1 Km Hg Mob, 11.4 kb	Mobilizable suicide plasmid for cloning out gene(s) adjacent to a Tn*5* insertion	C. Voisard, unpublished; Fig. 2

[a] Antibiotic resistance determinants are abbreviated as follows:
Ap, ampicillin; Hg, mercury; Km, kanamycin; Tc, tetracycline. Rep(ts), temperature-sensitive replication due to a mutation in the *trfA* gene. Mob, mobilization, Tra, conjugative transfer.

Plasmid transfer to strain CHA0 can be achieved by standard $CaCl_2$-dependent transformation or electroporation, by the methods used for *P. aeruginosa* (58). Cosmid libraries established in the broad-host-range IncP vector pVK100 are mobilizable from *Escherichia coli* to strain CHA0 with the help of pME497, a tetracycline-sensitive derivative of pME305 (Table 2). Other transfer systems commonly used, e.g., *E. coli* S17-1 (59) and pRK2013 (60), are not suitable for this purpose because of the conjugation barrier to wild-type IncP plasmids mentioned above.

We have also used pVK100 for subcloning of restriction fragments and for performing complementation tests in strain CHA0 (13, 33, 61). IncQ vectors, whose copy numbers are higher than those of the IncP family, replicate in many bacterial species including *P. fluorescens* CHA0 (62). However, we have avoided using IncQ vectors in strain CHA0 because of instability problems encountered with several IncQ recombinant plasmids (our unpublished observations).

Gene replacement experiments can be carried out conveniently with ColE1-based suicide vectors such as pME3087 and pME3088 (Fig.1). These plasmids are mobilizable by a derepressed ColI1 plasmid (63), e.g. R64*drd*-11 or pLG221, and carry multiple cloning sites plus a good selective marker (tetracycline resistance). In general, gene replacement is done in two steps. Firstly, recombinant pME3087 (pME3088) derivatives are mobilized to strain CHA0 and integrated into its genome via a single homologous recombination event, with selection for tetracycline resistance. Secondly, excision of the vector via another cross-over can be obtained by enrichment for tetracycline-sensitive cells (33, 61, 64).

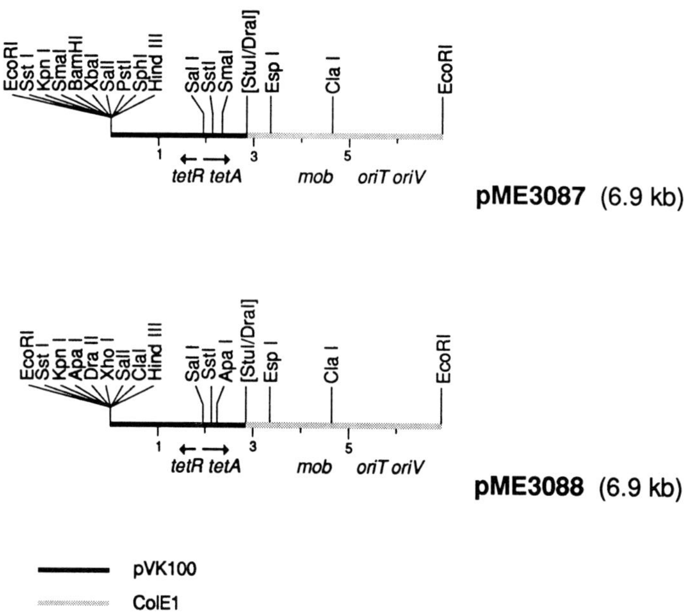

Fig. 1. Mobilizable suicide plasmids for gene exchange in *Pseudomonas* spp. These vectors consist of plasmid ColE1 (66), the tetracycline resistance determinant of pVK100 (65), and a polylinker from pMMB67EH (62; in pME3087) or from pBluescript (Stratagene; in pME3088). Restriction sites shown in brackets were lost during cloning; *tetR/tetA*, tetracycline resistance genes; *mob*, genes for mobilization by conjugative IncF, IncI or IncP plasmids; *oriT*, origin of transfer; *oriV*, origin of replication.

Another mobilizable ColE1 derivative, pME3049, has proved useful to clone the wild-type genes that correspond to those inactivated by a Tn*5* insertion in *P. fluorescens* CHA0. The rationale outlined in Fig. 2 consists of the following steps. Plasmid pME3049 carries the kanamycin resistance (Km) determinant of Tn*5*. After mobilization of pME3049 to a *P. fluorescens* mutant with a Tn*5* insertion in the *abc* gene(s), homologous recombination may occur between the Tn*5* segments, resulting in the chromosomal integration of pME3049 at the site of homology. The recovery of cells with a chromosomally integrated pME3049 is favored by selection for mercury resistance, the second resistance marker on pME3049 (Fig. 2). When chromosomal DNA from these cells is cut by *Eco*RI and self-ligated, a Km-resistant plasmid is produced that carries a flanking chromosomal fragment *(a)* and replicates in *E. coli* (Fig. 2). If this plasmid is then introduced into the wild-type CHA0 (with Km selection), homologous recombination in *a* can take place (Fig. 2). Subsequent restriction of chromosomal DNA with *Bgl*II and ligation allows the isolation of a new recombinant plasmid containing the intact *abc* gene(s) (Fig. 2). We have successfully applied this procedure to cloning genes involved in the regulation of Phl synthesis in strain CHA0 (our unpublished results).

Our standard approach in the genetic analysis of strain CHA0 has been to construct insertion or deletion mutants specifically affected in the production of one or several extracellular metabolites and to complement these mutants with the corresponding gene(s) cloned into pVK100 (65). Suppression of plant disease is then assessed for the wild-type, the mutant and the complemented mutant in a gnotobiotic system. The results and their interpretations are considered in section 6.2.4.

6.2.3 Gnotobiotic System

Disease suppression by fluorescent pseudomonads depends on complex interactions between the plant, the pathogen, the antagonistic biocontrol organism, and environmental factors. Therefore, studies on the mechanisms of disease suppression should be conducted under standardized conditions. This seems particularly important in the light of the working hypothesis that extracellular metabolites produced by the biocontrol agents are involved in disease suppression. The synthesis of siderophores and secondary metabolites depends strongly on external (biotic and abiotic) factors and on the growth phase of the producer organism. Standardization of environmental conditions (soil, humidity, temperature, light etc.) can be achieved in the laboratory. Weller et al. (67) have developed a standard test tube system to study the suppression of take-all of wheat by beneficial pseudomonads in natural soil. Other researchers have used gnotobiotic systems, i. e. conditions allowing plant growth in the presence of known biological components (68), to investigate interactions between the plant and beneficial bacteria (69, 70, 71). In our studies on *P. fluorescens* CHA0 we have established a gnotobiotic system that consists of a sterilized, artificial soil (composed of different quartz sand fractions, a pure clay mineral and a plant nutrient solution), sterile-grown plants, and defined numbers of pathogen and antagonist cells, in a flask plugged with cotton wool (10). This system has several advantages.

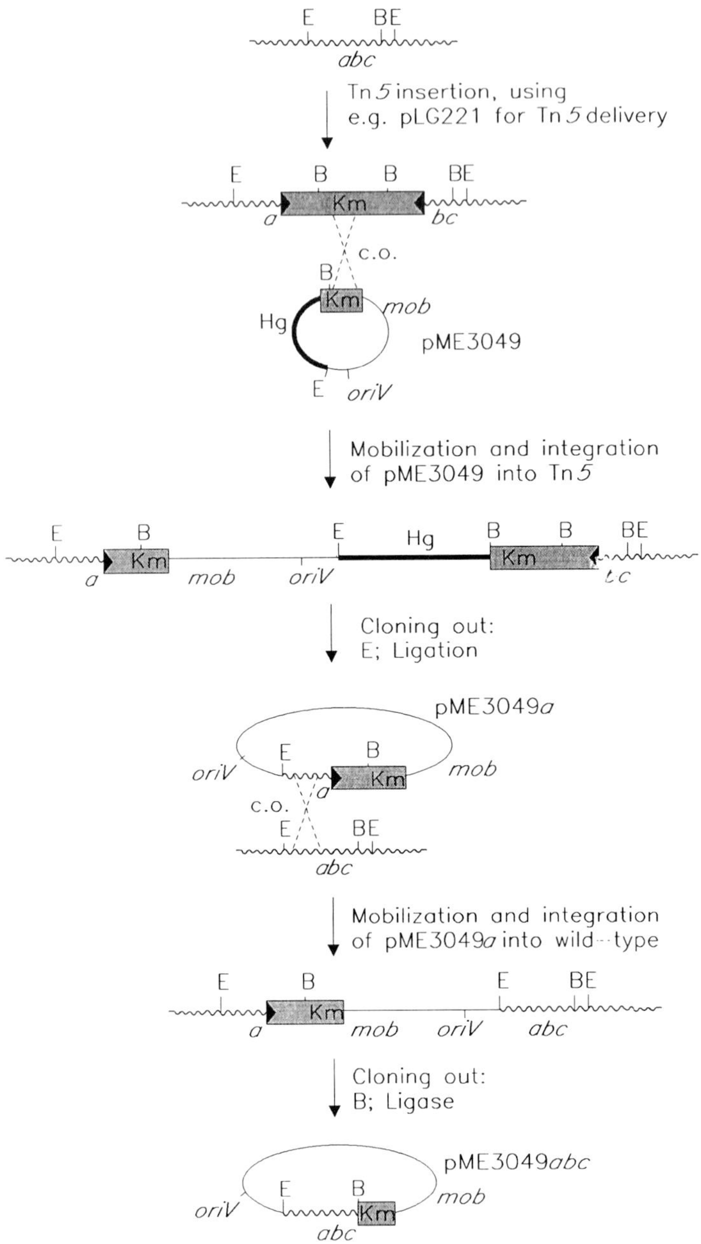

Fig. 2. Isolation of wild-type gene(s) corresponding to a Tn5 insertion mutation by *in vivo* cloning with the integrative vector pME3049. This plasmid is derived from ColE1 (–). It carries the kanamycin resistance determinant (Km) of Tn*5* and the mercury resistance determinant (Hg) of Tn*5-259* (56). B, *Bgl*II; E, *Eco*RI; *abc*, target gene(s) in the *Pseudomonas* genome; c.o., cross-over due to homologous recombination. Individual steps of the cloning procedure are described in the text.

(i) It consists of defined biological components and abiotic parameters, and these can be varied. (ii) Disease suppression can be measured quantitatively (in terms of plant weight) and qualitatively (in terms of disease index). Reproducible results are obtained with a range of plant-pathogen combinations (mentioned in the introduction). (iii) The system provides containment for genetically modified organisms. One drawback of the gnotobiotic system is that it excludes competition with other organisms and hence does not reflect field situations.

The gnotobiotic system can be used to evaluate some environmental factors influencing disease suppression. It is known that strain CHA0 suppresses disease effectively in soils collected from some field or regions but not from others (9, 72). In a gnotobiotic system containing vermiculite as clay mineral, strain CHA0 effectively controls black root rot. When vermiculite was replaced by illite, protection of tobacco was poor, but addition of Fe^{3+} to illite soil improved protection (10). The availability of iron is higher in vermiculite than in illite. Iron sufficiency favors cyanogenesis *in vitro* and may have the same effect in natural soil (10, 33). This mechanism could explain, in part, why disease suppression is influenced by the type of clay mineral.

6.2.4 Mutations Affecting Biocontrol Efficacy, Regulation of Secondary Metabolism, and some Caveats

Mutants of *P. fluorescens* CHA0 affected in the biosynthesis of pyoverdine, HCN, Phl, Plt and IAA have been obtained (Table 1). In the gnotobiotic system, mutations blocking HCN or Phl synthesis reduce the ability of strain CHA0 to suppress black root rot of tobacco; complementing plasmids largely restore disease suppression (13, 33). A Phl-negative mutant is also impaired in biocontrol of take-all and does not produce Phl in the rhizosphere of wheat (13, 73). We have interpreted these results to indicate that HCN and Phl each contribute, directly or indirectly, to the biocontrol properties of strain CHA0. By contrast, pyoverdine- or TSO-negative mutations had no significant effect (10, 31, 34). HCN production does not appear to be important for protection of wheat from take-all (32). Thus, suppression of different plant diseases by strain CHA0 may involve different mechanisms. Schippers et al. (74) have argued that HCN production by fluorescent pseudomonads may even have plant-deleterious effects.

Our observation that part of the biocontrol ability of strain CHA0 relies on antibiotic synthesis is supported by studies on several other plant-beneficial strains of fluorescent pseudomonads. *Pseudomonas* strains Q2-87 and F113 produce Phl and protect wheat from Ggt and sugar-beet from *Pythium ultimum*, respectively (75, 76). A Phl^- mutant of strain F113 affords a significantly reduced protection (77). Transfer of the complementing *phl* genes into heterologous *Pseudomonas* strains enables these strains to synthesize Phl and improves the biocontrol abilities of the strains (76, 77). The fluorescent pseudomonads 2-79 and 30-84 synthesize phenazine antibiotics. Phenazine-negative mutants show reduced capacity to suppress take-all due to Ggt (78, 79, 80, 81, 82). In another *P. fluorescens* strain, Hv37aR2, part of the suppression of *Pythium* infection on cotton can be attributed to the synthesis of the antibiotic oomycin A (17, 83). Although antibiotics produced by *Pseudomonas* species may not be crucial for the suppression of all root diseases (84) and although antibiotics should certainly not be viewed as exclusive factors in biological control, the current evidence does point to an involvement of antibiosis in many biocontrol systems.

We would like to mention two potential sources of errors in this kind of mutational analysis of disease suppression. Firstly, mutations affecting the synthesis of secondary metabolites can occur in regulatory genes which control a variety of cellular functions during restricted growth or stationary phase (61, 84). If the resulting mutants are defective in biocontrol, this may indicate, but need not prove, a role of antibiosis. There is always the possibility that inactivation of an antibiotic regulatory gene may result in additional unknown defects and that these defects may be responsible for reduced disease suppression. This problem can be circumvented to some extent if mutations are introduced into structural rather than regulatory genes for antibiotic synthesis. A second pitfall may occur even when a mutation in a structural gene blocks the production of an extracellular metabolite. The biosynthesis of the polyketide antibiotics Plt and Phl may require the same precursors (85). We have observed that mutations interfering with Phl synthesis in strain CHA0 lead to enhanced production of Plt and vice-versa (our unpublished results). Another example is presented by pyoverdine-negative mutants of strain CHA0; they overproduce salicylate (23). Thus, lack of one antibiotic or siderophore may be compensated by the increased synthesis of another product having similar properties. We would therefore like to conclude with a cautionary note. When a *Pseudomonas* mutant lacking a particular metabolite does not show altered biocontrol properties, this need not imply that this metabolite has no role in biocontrol. For instance, a pyoverdine-negative mutant of strain CHA0 protects tobacco and wheat with wild-type efficiency (Table 1). We infer from this result that pyoverdine does not have a paramount role in the suppression of black root rot and take-all, but iron competition is not ruled out as a mechanism because other siderophores such as salicylate might compensate for the lack of pyoverdine (10, 31).

Two genes have been identified which regulate the expression of secondary metabolites in strain CHA0. The **g**lobal **a**ctivator GacA, a response regulator in the family of two-component systems (86), is required for the synthesis of Phl, Plt, HCN and TSO (61). Moreover, GacA is involved in the expression of extracellular protease and phospholipase C (P. Sacherer, personal communication). A *gac*A mutant has a drastically reduced potential to suppress black root rot but still controls take-all by mechanisms that remain largely unknown at present (61). A second gene, *anr* encoding an **an**aerobic **r**egulator of transcription, the FNR-like protein ANR, controls cyanogenesis in *P. aeruginosa* PAO (87) and *P. fluorescens* CHA0 (our unpublished results). An *anr* mutant of strain CHA0 lacks HCN but synthesizes the antibiotics in wild-type amounts. In water-logged artificial soil, i. e. under conditions of poor oxygen supply, this mutant has a reduced ability to protect tobacco from *T. basicola* (our unpublished results). This finding is in agreement with the concept that HCN is involved in the suppression of black root rot.

Regulation of secondary metabolism by environmental factors is documented for fluorescent pseudomonads in some cases. For instance, HCN is optimally produced under conditions of oxygen limitation and in the presence of glycine and Fe (33, 88). Phl synthesis in strain F113 is stimulated by sucrose and growth on solid surfaces (75). High levels of Plt are produced in gluconate medium, and accumulation of phenazine-1-carboxylic acid is improved when Zn^{2+} or MoO_4^{2-} are provided (89, 90). However, there is no systematic investigation on how chemical signals – in particular root exudates and clay minerals – influence secondary metabolism in biocontrol pseudomonads.

Identification and cloning of genes that encode relevant traits opens the possibility to design fluorescent pseudomonads with improved or more reliable biocontrol activity (76, 77, 91, 92). Chromosomal insertion of the cloned *hcn* genes from strain CHA0 into *P. fluorescens* P3, a noncyanogenic strain that gives poor protection of tobacco, renders strain P3 Hcn^+ and improves its capacity to suppress black root rot (33). In another experiment, we have constructed a derivative of strain CHA0 that produces enhanced amounts of the antibiotics Phl and Plt *in vitro* and in the rhizosphere (14, 15). In disease suppression tests involving different plant – pathogen combinations, the antibiotic overproducer showed either improved plant protection or was inhibitory to some plants, presumably because of a toxic effect of the antibiotics on these plants (14, 15). Thus, genetically manipulated biocontrol strains should be tested very carefully before an eventual large-scale application.

6.2.5 Induced Systemic Resistance in Plants

Pseudomonads may be effective biocontrol agents because they antagonize a pathogen directly – e.g., by competition for nutrients or antibiosis – or indirectly, by activating plant defence mechanisms. Strain CHA0 colonizes root surfaces but has also been found in the root cortex (11). Such close association with roots may have effects on plant metabolism. Tobacco roots grown in the presence of strain CHA0 show increased root hair formation (11). In recent studies we have addressed the question whether root colonization by strain CHA0 might also activate defence mechanisms in the plant. There are precedents for this hypothesis. The fluorescent *Pseudomonas* sp. strain WCS417r colonizes the roots of carnation and protects the plant from Fusarium wilt. Disease was significantly reduced when roots were bacterized one week prior to stem-inoculation with conidia of *Fusarium oxysporum* f.sp. *dianthi*. The spacial separation between the beneficial bacterium and the pathogen practically excludes competition or direct antagonism as mechanisms (93). Heat-killed cells or lipopolysaccharides (LPS) of strain WCS417r applied to roots were also protective (94). These authors therefore suggested that strain WCS417r induces resistance in carnation and LPS could be an inducing signal.

Seed treatment with plant-growth promoting rhizobacteria has been found to reduce the severity of cucumber anthracnose, a leaf disease caused by *Colletotrichum orbiculare* (95). In another study, fewer halo blight lesions (due to *P. syringae* pv. *phaseolicola*) were observed on beans when seeds had been bacterized with a plant-beneficial strain of *P. fluorescens* (96). Certain fluorescent pseudomonads, after root inoculation, induce mRNAs that encode the pathogenesis-related protein PR1a in bean leaves (97). Taken together, these observations suggest that certain root-colonizing bacteria may induce systemic resistance in plants. However, it has not been ruled out that some bacterial metabolites (e.g., antibiotics) could be taken up and translocated by the plants; at the infection site, these metabolites could exert an antagonistic effect, similar to the systemic protection obtained with certain chemical pesticides.

We have tested whether root-colonizing strain CHA0 induces systemic resistance of tobacco against tobacco necrosis virus (TNV) applied to leaves. Strain CHA0 (10^7 cells/g of soil) were added to autoclaved soil three days before planting four-week-old *Nicotiana tabacum* cultivar Xanthi nc. Six weeks later the youngest fully developed leaf was inoculated with TNV. One week after this challenge inoculation the leaves of plants grown in the presence of strain CHA0 showed lesions that were reduced in numbers (by about 60%) and in size (by about 20%), compared to leaves of nonbacterized plants. Inoculation of tobacco with a rifampicin-resistant mutant of strain CHA0 gave the same results and no bacteria could be reisolated from the stem or leaves on selective medium containing rifampicin (M. Maurhofer, unpublished data). We suggest that strain CHA0 is able to induce some systemic resistance in the plant but we do not know the signal(s) that may trigger the induction. Since exogenous salicylate can act as an inducer of systemic resistance (29), it would be interesting to isolate and test a salicylate-negative mutant of strain CHA0.

6.2.6 Genetic Instability of Strain CHA0: Effects on Secondary Metabolism and Biological Control

The biocontrol properties of a given strain should not be lost easily during handling in the laboratory and quality control procedures should be available that ensure the maintenance of traits essential for biocontrol activity. It is a well-known fact that under laboratory conditions some soil bacteria (especially *Streptomyces* spp.) can lose the ability to produce secondary metabolites and some fungal pathogens (e.g., Ggt) tend to become hypovirulent (98, 99). In *Streptomyces* spp. particular phenotypes can be lost at high frequencies due to spontaneous chromosomal deletions or DNA amplifications (99, 100). A related phenomenon has been observed in many commonly used laboratory strains of *E. coli* K-12: They carry a point mutation in the *rpoS* gene encoding the stationary-phase sigma factor. The *rpoS* mutants survive less well under harsh environmental conditions than does the wild-type and they may be unable to express virulence determinants (101, 102). There are also indications of genetic instability in *Pseudomonas* spp. but the underlying molecular mechanisms are poorly understood (7, 103, 104).

Strain CHA0 shows signs of pronounced genetic instability during prolonged exposure to rich media under laboratory conditions, but in the absence of any obvious mutagen. After several days or weeks of incubation, CHA0 mutants can be isolated that lack TSO activity and/or the production of secondary metabolites (105). Moreover, several pleiotropic mutants are also defective in extracellular protease (P. Sacherer, personal communication). These initial observations have recently been extended. Fifty-three independent cultures of strain CHA0 were incubated in nutrient yeast broth at 30 °C, with or without shaking, for 3 to 6 weeks. Fifty single colonies obtained from each culture were screened for the marker enzyme TSO, extracellular protease activity, and auxotrophy. In this way, one auxotrophic mutant and 50 TSO^- and protease-negative mutants were obtained. These were further tested for the production of HCN and antibiotics. Thus far, we have characterized two pleiotropic mutants lacking TSO, protease, HCN, and antibiotics. All these phenotypes were restored by the introduction of the *gacA*$^+$ plasmid pME3066, suggesting that the spontaneous mutations affect the *gacA* gene. Four phenotypically similar mutants could not be complemented by pME3066 (our unpublished results). We therefore suppose that the expression of secondary metabolites and extracellular enzymes in strain CHA0 requires at least two, but possibly a large number of regulatory genes. In *Streptomyces coelicolor*, a cascade of regulatory elements controls antibiotic synthesis (106).

The loss of critical phenotypes due to laboratory maintenance of biocontrol strains is unsettling and accentuates the danger of a decline of biocontrol effectiveness when these strains are later applied in the field or greenhouse. Quality control is needed to maintain the effectiveness of biocontrol strains. With respect to strain CHA0, we can specify a minimum requirement for the suppression of some root diseases such as black root rot of tobacco: the bacterium should produce secondary metabolites (especially HCN and Phl). Since mutational defects in secondary metabolism often correlate with a lack of TSO expression, this enzyme may be a suitable cytochemical marker to monitor genetic instability in strain CHA0. A limitation of this approach is that some bacterial phenotypes which are important to biological control (e. g., root colonization [4, 78]), cannot be assayed easily. A rapid and simple screening test for rhizosphere competence would be needed.

To minimize the incidence of spontaneous mutations affecting the biocontrol properties of strain CHA0, we routinely stock this bacterium and its derivatives at −80 °C in 50% glycerol and we do not keep the bacteria in nutrient media at 4 to 30 °C for more than 3 to 4 days. When a storage vial is removed from the freezer, the culture is thawed on the surface to allow plating of the bacteria. Repeated freezing and thawing of cultures is avoided as this would result in a loss of viability. We recommend that biocontrol bacteria newly isolated from soil should also be lyophilized immediately for safe storage.

6.3 Environmental Impact of Bacterial Inoculants

Biological control of root diseases with wild-type or genetically modified microorganisms requires the deliberate release of these microorganisms in large numbers into soil ecosystems. Although plant-beneficial, fluorescent pseudomonads have not been observed to cause damaging effects on the environment, it appears appropriate to determine the ecological competence (e.g., root colonization, survival in the rhizosphere and in neighbouring soil) of released microorganisms, their dissemination (e.g., to ground water) and their impact on resident microbial populations (107).

6.3.1 Environmental Monitoring

Methods are needed to detect released pseudomonads in the environment and to identify the strains reliably (108, 109). Selective plating still seems to be the most sensitive method to recover a particular strain from soil. We have marked strain CHA0 with spontaneous rifampicin or nalidixic acid resistance. Plating on amended S1 medium, which contains sodium lauroyl sarcosine, trimethoprim (110) and rifampicin or nalidixic acid, allows us to detect one viable cell per g of soil. Following isolation on selective media, strain CHA0 can be identified by the following criteria. (i) Specific polyclonal antibodies directed against the cell wall of strains CHA0 have been labeled with fluorescein-isothiocyanate. They enable the bacterium to be detected by immunofluorescence microscopy. This technique has also been used to visualize root colonization by strain CHA0 (11). (ii) Strain CHA0 has a typical profile of secondary metabolites which can be monitored *in vitro* (Table 1). (iii) Southern hybridization of genomic DNA with *phl* genes probes (13, 76) has shown that strain CHA0 gives a characteristic restriction pattern. Similar (but not identical) patterns have been found in other Phl producing pseudomonads isolated from Swiss and U.S. soils (C. Keel, L. Thomashow & D. Weller, unpublished results). In future work, we plan to expand the range of DNA probes and to develop immunofluorescence techniques to monitor viable but non-culturable cells (111, 112).

6.3.2 Microcosms

Soil microcosms have been designed for laboratory use to study the persistence and translocation of introduced microorganisms under conditions that are close to those in natural environments (113, 114, 115, 116). We have investigated the survival and transport of strain CHA0 in undisturbed soil columns, using a modification of the ASTM (American Society for Testing and Materials) standard soil-core microcosm, which was developed to investigate the percolation of chemicals (117). A sieved surface layer soil is packed onto an intact deeper-layer soil core and placed in a PVC pipe; small gaps between the soil core and the pipe are filled with paraffin to eliminate boundary effects. The top of the column is covered with a thin layer of bacterized soil and wheat is planted (Fig.3). Weekly rainfall is simulated by sprinkling appropriate amounts of water. In our preliminary experiments involving nonsterile soil, introduced strain CHA0 could be detected in different depths of the soil core and in the percolation water collected at the bottom of the column. Cell num-

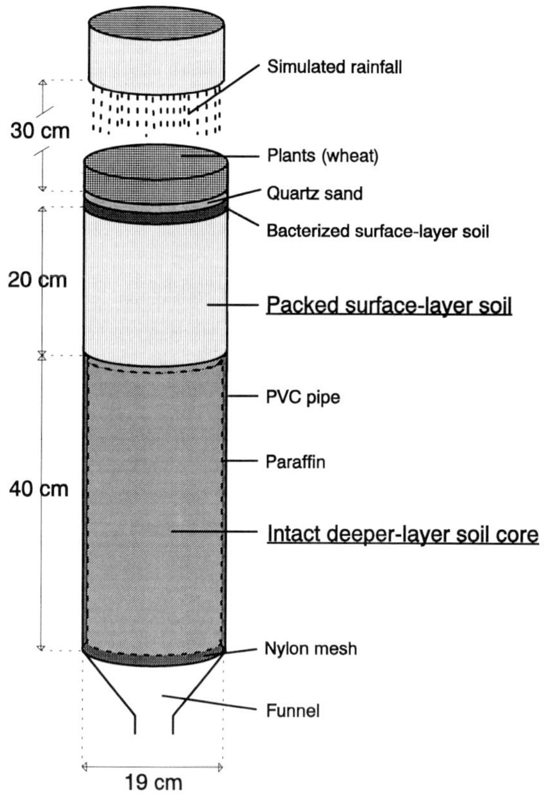

Fig. 3. Microcosm simulating a soil horizon for studies on the fate of introduced microorganisms. This system is based on the ASTM standard soil-core microcosm. The soil-core is taken intact from a natural environment.

bers declined constantly over several weeks. Compared to the wild-type, a *gacA* mutant was less persistent in soil outside the rhizosphere but had no competitive disadvantage when colonizing roots. We next intend to monitor the fate of introduced bacteria, with and without genetic modification, for several months.

6.4 Potential Applications

Under laboratory and experimental greenhouse conditions *P. fluorescens* CHA0 and other biocontrol strains give reproducible and effective plant protection against a range of major root pathogens. Successful applications of beneficial rhizobacteria under field conditions are well documented but the performance of introduced biocontrol strains tends to be highly variable (6, 7, 118). In most, but not all, field experiments conducted between 1986 and 1990 strain CHA0 or its nalidixic acid resistant mutant CO3 significantly increased plant yields when applied to field plots naturally or artificially infested with Ggt. However, the same strains did not protect tobacco in field plots artificially infested with *T. basicola* (11, 119; B. Wüthrich, personal communication).

Inconsistent biocontrol performance may have several reasons (7). (i) Loss of ecological competence: We have pointed out the genetic instability of strain CHA0 under certain laboratory conditions. Spontaneous, pleiotropic mutations in *gacA* or other genes controlling secondary metabolism may reduce the bacterium's potential to suppress disease. (ii) Local influence of environmental factors: The variation of protection obtained with strain CHA0 in different soils has been attributed to variable expression of desirable biocontrol traits under different environmental conditions (9, 10, 72). Variable root colonization may be a further important cause of inconsistent performance (7, 78). (iii) Occurrence of non-target pathogens, which are insensitive to the biocontrol agent, can lead to failure (7, 120).

An approach to overcome the problem of inconsistent performance in the field is to apply *mixtures* of different biocontrol strains. Combinations of different *Pseudomonas* strains improve the protection of wheat against take-all in field plots (7, 121), and the combination a fluorescent *Pseudomonas* strain with a non-pathogenic *Fusarium* strain has been used to control Fusarium wilt (122). A CHA0-like *P. fluorescens* strain, which was isolated with CHA0 from the same site, gives significantly improved suppression of Fusarium wilt of tomato when combined with the non-pathogenic *Fusarium* strain Fo47 (123). A mixture of another *Pseudomonas* strain (Pf36) and strain Fo47 improved the yields of first-quality tomatoes by about 20% and reduced the incidence of disease, in a soil naturally infested with *F. oxysporum* f.sp. *lycopersici* under commercial greenhouse conditions. Application of the non-pathogenic *Fusarium* strain alone had no effect (J. Fuchs, personal communication).

6.5 Conclusion

Suppression of root diseases by fluorescent pseudomonads depends on multiple biotic and abiotic factors. The relative importance of siderophores and certain antibiotics in disease suppression has been assessed for selected biocontrol strains and host-pathogen interactions (10, 32, 79, 124, 125). Antibiotic-negative mutants of *Pseudomonas* spp. are less suppressive in many, but not all, host-pathogen systems tested (61, 77, 83, 84, 125). One remaining challenge will be to find new bacterial factors involved in the biocontrol of those diseases that are clearly suppressed by antibiotic-negative strains. In other host-pathogen systems, which are susceptible to secondary metabolites of *Pseudomonas* spp., the challenge will be to see whether there are cause-effect relationships between environmental signals, bacterial cell densities, production of secondary metabolites, and biocontrol effectiveness. Finally, it will be interesting to explore the molecular mechanisms governing genetic instability of secondary metabolism and biocontrol efficacy in *Pseudomonas* spp.

Acknowledgements

We thank Moshe Flaishman, Jacques Fuchs, Valeria Gaia, Andreas Natsch, Hanspeter Pfirter, Priska Sacherer, Josef Troxler, Christine von Schroetter, Gordon Stewart, Paul Williams, and Beat Wüthrich for contributing and communicating unpublished results and Anita Fischer for providing secretarial assistance. Support from the Schweizerische Nationalfonds (projects 31-28570.90 and 31-32473.91; priority programme biotechnology 5002-35142) and from the U.S. National Science Foundation (INT-9102689; to C.T.B.) is gratefully acknowledged.

6.6 References

1. Baker, K.F. Evolving concepts of biological control of plant pathogens. Annu. Rev. Phytopathol. *25* (1987) 67-85.
2. Défago, G., and D. Haas. Pseudomonads as antagonists of soilborne plant pathogens: modes of action and genetic analysis. Soil Biochem. *6* (1990) 249-291.
3. Kloepper, J.W., R. Lifshitz, and M.N. Schroth. *Pseudomonas* inoculants to benefit plant production. ISI Atlas of Science: Animal and Plant Sciences: (1988) 60-64.
4. Lugtenberg, B.J.J., L.A. de Weger, and J.W. Bennett. Microbial stimulation of plant growth and protection from disease. Current Opinion in Biotechnology *2* (1991) 457-464.
5. O'Sullivan, D.J., and F. O'Gara. Traits of fluorescent *Pseudomonas* spp. involved in suppression of plant root pathogens. Microbiol. Rev. *56* (1992) 662-676.
6. Schippers, B., A.W. Bakker, and P.A.H.M. Bakker. Interactions of deleterious and beneficial rhizosphere microorganisms and the effect of cropping practices. Annu. Rev. Phytopathol. *25* (1987) 339-358.
7. Weller, D.M. Biological control of soilborne plant pathogens in the rhizosphere with bacteria. Annu. Rev. Phytopathol. *26* (1988) 379-407.
8. Zablotowicz, R.M., C. M. Press, N. Lyng, G.L. Brown, and J.W. Kloepper. Compatibility of plant growth promoting rhizobacterial strains with agrichemicals applied to seed. Can. J. Microbiol. *38* (1992) 45-50.
9. Stutz, E., G. Défago, and H. Kern. Naturally occurring fluorescent pseudomonads involved in suppression of black root rot of tobacco. Phytopathology *76* (1986) 181-185.
10. Keel, C., C. Voisard, C.H. Berling, G. Kahr, and G. Défago. Iron sufficiency, a prerequisite for suppression of tobacco black root rot by *Pseudomonas fluorescens* strain CHA0 under gnotobiotic conditions. Phytopathology *79* (1989) 584-589.
11. Défago, G., C.H. Berling, U. Burger, D. Haas, G. Kahr, C. Keel, C. Voisard, P. Wirthner, and B. Wüthrich. Suppression of black root rot of tobacco and other root diseases by strains of *Pseudomonas fluorescens*: potential applications and mechanisms. In: D. Hornby, R.J. Cook, Y. Henis, W.H. Ko, A.D. Rovira, B. Schippers and P.R. Scott (eds.), Biological control of soil-borne plant pathogens. CAB International, 1990, p. 93-108.
12. Keel, C., P. Wirthner, T. Oberhänsli, C. Voisard, U. Burger, D. Haas, and G. Défago. Pseudomonads as antagonists of plant pathogens in the rhizosphere: role of the antibiotic 2,4-diacetylphloroglucinol in the suppression of black root rot of tobacco. Symbiosis *9* (1990) 327-341.
13. Keel, C., U. Schnider, M. Maurhofer, C. Voisard, J. Laville, U. Burger, P. Wirthner, D. Haas, and G. Défago. Suppression of root diseases by *Pseudomonas fluorescens* CHA0: importance of the bacterial secondary metabolite 2,4-diacetylphloroglucinol. Mol. Plant-Microbe Inter. *5* (1992) 4-13.
14. Maurhofer, M., C. Keel, U. Schnider, C. Voisard, D. Haas, and G. Défago. Does enhanced antibiotic production in *Pseudomonas fluorescens* strain CHA0 improve its disease suppressive capacity? In: C. Keel, B. Koller, and G. Défago (eds.), Plant growth-promoting rhizobacteria – Progress and prospects. IOBC/WPRS Bulletin 14. 1991, p. 201-202.

15. Maurhofer, M., C. Keel, U. Schnider, C. Voisard, D. Haas, and G. Défago. Influence of enhanced antibiotic production in *Pseudomonas fluorescens* strain CHA0 on its disease suppressive capacity. Phytopathology *82* (1992) 190–195.
16. Fravel, D.R. Role of antibiosis in the biocontrol of plant diseases. Annu. Rev. Phytopathol. *26* (1988) 75–91.
17. Gutterson, N. Microbial fungicides: recent approaches to elucidating mechanisms. Crit. Rev. Biotechnol. *10* (1990) 69–91.
18. Handelsman, J., and J.L. Parke. Mechanisms in biocontrol of soilborne plant pathogens. In: T. Kosuge and E. Nester (eds.), Plant-microbe interactions, vol. III, McGraw Hill, New York, 1989, p. 27–61.
19. Leong, J. Siderophores: their biochemistry and possible role in the biocontrol of plant pathogens. Annu. Rev. Phytopathol. *24* (1986) 187–209.
20. Loper, J.E., and J.S. Buyer. Siderophores in microbial interactions on plant surfaces. Mol. Plant-Microbe Inter. *4* (1991) 5–13.
21. Schroth, M.N., and J.G. Hancock. Disease-suppressive soil and root-colonizing bacteria. Science *216* (1982) 1376–1381.
22. Ahl, P., C. Voisard, and G. Défago. Iron bound-siderophores, cyanide, and antibiotics involved in suppression of *Thielaviopsis basicola* by a *Pseudomonas fluorescens* strain. J. Phytopathol. *116* (1986) 121–134.
23. Meyer, J.M., P. Azelvandre, and C. Georges. Iron metabolism in *Pseudomonas*: salicylic acid, a siderophore of *Pseudomonas fluorescens* CHA0. BioFactors *4* (1992) 23–27.
24. Visca, P., G. Colotti, L. Serino, D. Verzili, N. Orsi, and E. Chiancone. Metal regulation of siderophore synthesis in *Pseudomonas aeruginosa* and functional effects of siderophore-metal complexes. Appl. Environ. Microbiol. *58* (1992) 2886–2893.
25. Ankenbauer, R.G., and C.D. Cox. Isolation and characterization of *Pseudomonas aeruginosa* mutants requiring salicylic acid for pyochelin biosynthesis. J. Bacteriol. *170* (1988) 5364–5367.
26. Ratledge, C., L.P. Macham, K.A. Brown, and B.J. Marshall. Iron transport in *Mycobacterium smegmatis*: A restricted role for salicylic acid in the extracellular environment. Biochim. Biophys. Acta *372* (1974) 39–51.
27. Enyedi, A.J., N. Yalpani, P. Silverman, and I. Raskin. Signal molecules in systemic plant resistance to pathogens and pests. Cell *70* (1992) 879–886.
28. Raskin, I. Salicylate, a new plant hormone. Plant Physiol. *99* (1992) 799–803.
29. Raskin, I. Role of salicylic acid in plants. Annu. Rev. Plant Mol. Biol. *43* (1992) 439–463.
30. Cuppels, D.A., R.D. Stipanovic, A. Stoessl, and J.B. Stothers. The constitution and properties of a pyochelin-zinc complex. Can. J. Chem. *65* (1987) 2126–2130.
31. Keel, C., and G. Défago. The fluorescent siderophore of *Pseudomonas fluorescens* strain CHA0 has no effect on the suppression of root diseases of wheat. In: C. Keel, B. Koller, and G. Défago (eds.), Plant growth-promoting rhizobacteria – Progress and prospects. IOBC/WPRS Bulletin 14. 1991, p. 136–142.
32. Haas, D., C. Keel, J. Laville, M. Maurhofer, T. Oberhänsli, U. Schnider, C. Voisard, B. Wüthrich, and G. Défago. Secondary metabolites of *Pseudomonas fluorescens* strain CHA0 involved in the suppression of root diseases. In: H. Hennecke and D.P.S. Verma (eds.), Advances in molecular genetics of plant-microbe interactions, Vol. I. Kluwer Academic Publishers, Dordrecht, 1991, p. 450–456.
33. Voisard, C., C. Keel, D. Haas, and G. Défago. Cyanide production by *Pseudomonas fluorescens* helps suppress black root rot of tobacco under gnotobiotic conditions. EMBO J. *8* (1989) 351–358.
34. Oberhänsli, T., G. Défago, and D. Haas. Indole-3-acetic acid (IAA) synthesis in the biocontrol strain CHA0 of *Pseudomonas fluorescens*: role of tryptophan side chain oxidase. J. Gen. Microbiol. *137* (1991) 2273–2279.
35. Bainton, N.J., B.W. Bycroft, S.R. Chhabra, P. Stead, L. Gledhill, P.J. Hill, C.E.D. Rees, M.K. Winson, G.P.C. Salmond, C.S.A.B. Stewart, and P. Williams. A general role for the *lux* autoinducer in bacterial cell signalling: control of antibiotic biosynthesis in *Erwinia*. Gene *116* (1992) 87–91.
36. Way, J.L., P. Leung, E. Cannon, R. Morgan, C. Tamulinas, J.L. Way, L. Baxter, A. Nagi, and C. Chui. The mechanism of cyanide intoxication and its antagonism. In: D. Evered and S. Harnett (eds.), Cyanide compounds in biology. J. Wiley, Chichester, 1988, p. 232–243.
37. Akimenko, V.K., S.M. Trutko, A.G. Medentsev, and V.P. Korobov. Distribution of cyanide-resistant respiration among yeasts and bacteria and its relation to oversynthesis of metabolites. Arch. Microbiol. *136* (1983) 234–241.

38. Matsushita, K., M. Yamada, E. Shinagawa, O. Adachi, and M. Ameyama. Membrane-bound respiratory chain of *Pseudomonas aeruginosa* grown aerobically. A KCN-insensitive alternate oxidase chain and its energetics. J. Biochem. *93* (1983) 1137-1144.
39. Zannoni, D. The respiratory chains of pathogenic pseudomonads. Biochim. Biophys. Acta *975* (1989) 299-316.
40. Barbieri, P., T. Zanelli, E. Galli, and G. Zanetti. Wheat inoculation with *Azospirillum brasilense* Sp6 and some mutants altered in nitrogen fixation and indole-3-acetic acid production. FEMS Microbiol. Lett. *36* (1986) 87-90.
41. Crozier, A., P. Arruda, J.M. Jasmin, A.M. Monteiro, and G. Sandberg. Analysis of indole-3-acetic acid and related indoles in culture medium from *Azospirillum lipoferum* and *Azospirillum brasilense*. Appl. Environ. Microbiol. *54* (1988) 2833-2837.
42. Défago, G. 2,4-Diacetylphloroglucinol, a promising compound in biocontrol. Plant Pathology (1993) (in press).
43. Dutrecq, A., P. Debras, J. Stevaux, and M. Marlier. Activity of 2,4-diacetylphloroglucinol isolated from a strain of *Pseudomonas fluorescens* to *Gaeumannomyces graminis* var. *tritici*. In: A.B.R. Beemster, G.J. Bollen, M. Gerlagh, M.A. Ruissen, B. Schippers, and A. Tempel (ed.), Biotic interactions and soil-borne diseases. Elsevier Science Publishers, Amsterdam 1991, p. 252-257.
44. Homma, Y., Z. Sato, F. Hirayama, K. Konno, H. Shirahama, and T. Suzui. Production of antibiotics by *Pseudomonas cepacia* as an agent for biological control of soilborne plant pathogens. Soil Biol. Biochem. *21* (1989) 723-728.
45. Howell, C.R., and R.D. Stipanovic. Control of *Rhizoctonia solani* on cotton seedlings with *Pseudomonas fluorescens* and with an antibiotic produced by the bacterium. Phytopathology *69* (1979) 480-482.
46. Howell, C.R., and R.D. Stipanovic. Suppression of *Pythium ultimum*-induced damping-off of cotton seedlings by *Pseudomonas fluorescens* and its antibiotic, pyoluteorin. Phytopathology *70* (1980) 712-715.
47. Kiprianova, E.A., and Smirnov, V.V. *Pseudomonas fluorescens*, producer of antibiotic compounds. Antibiotiki *26* (1981) 135-143.
48. Ohmori, T., S. Hagiwara, A. Ueda, Y. Minoda, and K. Yamada. Production of pyoluteorin and its derivatives from *n*-paraffin by *Pseudomonas aeruginosa* S10B2. Agric. Biol. Chem. *42* (1978) 2031-2036.
49. Dunlap, P.V., and A. Kuo. Cell density-dependent modulation of the *Vibrio fischeri* luminescence system in the absence of autoinducer and LuxR protein. J. Bacteriol. *174* (1992) 2440-2448.
50. Meighen, E.A. Molecular biology of bacterial bioluminescence. Microbiol. Rev. *55* (1991) 123-142.
51. Stewart, G.S.A.B., and P. Williams. *Lux* genes and the applications of bacterial bioluminescence. J. Gen. Microbiol. *138* (1992) 1289-1300.
52. Williams, P., N.J. Bainton, S. Swift, S.R. Chhabra, M.K. Winson, G.S.A.B. Stewart, G.P.C. Salmond, and B.W. Bycroft. Small molecule-mediated density-dependent control of gene expression in prokaryotes: bioluminescence and the biosynthesis of carbapenem antibiotics. FEMS Microbiol. Lett. *100* (1992) 161-168.
53. Bainton, N.J., P. Stead, S.R. Chhabra, B.W. Bycroft, G.P.C. Salmond, G.S.A.B. Stewart, and P. Williams. *N*-(3-oxohexanoyl)-L-homoserine lactone regulates carbapenem antibiotic production in *Erwinia carotovora*. Biochem. J. *288* (1992) 997-1004.
54. Beppu, T. Secondary metabolites as chemical signals for cellular differentiation. Gene *115* (1992) 159-165.
55. Boulnois, G.J., J.M. Varley, G.S. Sharpe, and C.F. Franklin. Transposon donor plasmids, based on Collb-P9, for use in *Pseudomonas putida* and a variety of other gram negative bacteria. Mol. Gen. Genet. *200* (1985) 65-67.
56. Voisard, C., M. Rella, and D. Haas. Conjugative transfer of plasmid RP1 to soil isolates of *Pseudomonas fluorescens* is facilitated by certain large RP1 deletions. FEMS Microbiol. Lett. *55* (1988) 9-14.
57. Ubben, D., and R. Schmitt. Tn*1721* derivatives for transposon mutagenesis, restriction mapping and nucleotide sequence analysis. Gene *41* (1986) 145-152.
58. Farinha, M.A., and A.M. Kropinski. High efficiency electroporation of *Pseudomonas aeruginosa* using frozen cell suspensions. FEMS Microbiol. Lett. *70* (1990) 221-226.

59. Simon, R., U. Priefer, and A. Pühler. Vector plasmids for in-vivo and in-vitro manipulations of gram-negative bacteria. In: A. Pühler (ed.), Molecular genetics of the bacteria-plant interaction. Springer-Verlag, Berlin, 1983, p. 98–106.
60. Figurski, D.H. and D.R. Helinski. Replication of an origin-containing derivative of plasmid RK2 dependent on a plasmid function provided in *trans*. Proc. Natl. Acad. Sci. USA *76* (1979) 1648–1652.
61. Laville, J., C. Voisard, C. Keel, M. Maurhofer, G. Défago, and D. Haas. Global control in *Pseudomonas fluorescens* mediating antibiotic synthesis and suppression of black root rot of tobacco. Proc. Natl. Acad. Sci. USA *89* (1992) 1562–1566.
62. Morales, V., M.M. Bagdasarian, and M. Bagdasarian. Promiscuous plasmids of the IncQ group: mode of replication and use for gene cloning in gram-negative bacteria. In: S. Silver, A.M. Chakrabarty, B. Iglewski, and S. Kaplan (eds.), *Pseudomonas*: Biotransformations, pathogenesis, and evolving biotechonology. American Society for Microbiology, Washington, D.C., 1990, p. 229–241.
63. Finnegan, J., and D. Sherratt. Plasmid ColE1 conjugal mobility: The nature of *bom*, a region required in *cis* for transfer. Mol. Gen. Genet. *185* (1982) 344–351.
64. Haas, D., J. Watson, R. Krieg, and T. Leisinger. Isolation of an Hfr donor of *Pseudomonas aeruginosa* PAO by insertion of the plasmid RP1 into the tryptophan synthase gene. Mol. Gen. Genet. *182* (1981) 240–244.
65. Knauf, V.C., and E.W. Nester. Wide host range cloning vectors: A cosmid clone bank of an Agrobacterium Ti plasmid. Plasmid *8* (1982) 45–54.
66. Chan, P.T., H. Ohmori, J. Tomizawa, and J. Lebowitz. Nucleotide sequence and gene organization of ColE1 DNA. J. Biol. Chem. *260* (1985) 8925–8935.
67. Weller, D. M., B.-X. Zhang, and R.J. Cook. Application of a rapid screening test for selection of bacteria suppressive to take-all of wheat. Plant Disease *69* (1985) 710–713.
68. Hale, M. G., D.L. Lindsey, and K.M. Hameed. Gnotobiotic culture of plants and related research. Bot. Rev. *39* (1973) 261–273.
69. Bennett, R.A., and Lynch, J. M. Bacterial growth and development in the rhizosphere of gnotobiotic cereal plants. J. Gen. Microbiol. *125* (1981) 95–102.
70. Kloepper, J.W., and M.N. Schroth. Plant growth-promoting rhizobacteria and plant growth under gnotobiotic conditions. Phytopathology *71* (1981) 642–644.
71. Lifshitz, R., J.W. Kloepper, M. Kozlowski, C. Simonson, J. Carlson, E.M. Tipping, and I. Zaleska. Growth promotion of canola (rapeseed) seedlings by a strain of *Pseudomonas putida* under gnotobiotic conditions. Can. J. Microbiol. *33* (1987) 390–395.
72. Stutz, E., G. Kahr, and G. Défago. Clays involved in suppression of tobacco black root rot by a strain of *Pseudomonas fluorescens*. Soil Biol. Biochem. *21* (1989) 361–366.
73. Keel, C., M. Maurhofer, T. Oberhänsli, C. Voisard, D. Haas, and G. Défago. Role of 2,4-diacetylphloroglucinol in the suppression of take-all of wheat by a strain of *Pseudomonas fluorescens*. In: A.B.R. Beemster, G.J. Bollen, M. Gerlagh, M.A. Ruissen, B. Schippers and A. Tempel (eds.), Biotic interactions and soil-borne diseases. Elsevier, Amsterdam, 1991, p. 335–339.
74. Schippers, B., A.W. Bakker, P.A.H.M. Bakker, and R. Van Peer. Beneficial and deleterious effects of HCN producing pseudomonads on rhizospere interactions. In: D.L. Keister, and P.B. Cregan (eds.), The rhizosphere and plant growth. Kluwer Academic Publishers, Dordrecht, 1991, p. 211–219.
75. Shanahan, P., D.J. O'Sullivan, P. Simpson, J. D. Glennon, and F. O'Gara. Isolation of 2,4-diacetylphloroglucinol from a fluorescent pseudomonad and investigation of physiological parameters influencing its production. Appl. Environ. Microbiol. *58* (1992) 353–358.
76. Vincent, M. N., L.A. Harrison, J.M. Brackin, P. Kovacevich, P. Mukerji, D.M. Weller, E.A. Pierson. Genetic analysis of the antifungal activity of a soilborne *Pseudomonas aureofaciens* strain. Appl. Environ. Microbiol. *57* (1991) 2928–2934.
77. Fenton, A.M., P.M. Stephens, J. Crowley, M. O'Callaghan, and F. O'Gara. Exploitation of gene(s) involved in 2,4-diacetylphloroglucinol biosynthesis to confer a new biocontrol capability to a *Pseudomonas* strain. Appl. Environ. Microbiol. *58* (1992) 3873–3878.
78. Bull, C.T., D.M. Weller, and L.S. Thomashow. Relationship between root colonization and suppression of *Gaeumannomyces graminis* var. *tritici* by *Pseudomonas fluorescens* strain 2–79. Phytopathology *81* (1991) 954–959.
79. Hamdan, H., D.M. Weller, and L.S. Thomashow. Relative importance of fluorescent siderophores and other factors in biological control of *Gaeumannomyces graminis* var. *tritici* by *Pseudomonas fluorescens* 2–79 and M4–80R. Appl. Environ. Microbiol. *57* (1991) 3270–3277.

80. Pierson III, L.S., and L.S. Thomashow. Cloning and heterologous expression of the phenazine biosynthetic locus from *Pseudomonas aureofaciens* 30-84. Mol. Plant-Microbe Inter. *5* (1992) 330-339.
81. Thomashow, LS., and D.M. Weller. Role of a phenazine antibiotic from *Pseudomonas fluorescens* in biological control of *Gaeumannomyces graminis* var. *tritici*. J. Bacteriol. *170* (1988) 3499-3508.
82. Thomashow, LS., D.M. Weller, R.F. Bonsall, and L.S. Pierson III. Production of the antibiotic phenazine-1-carboxylic acid by fluorescent *Pseudomonas* species in the rhizosphere of wheat. Appl. Environ. Microbiol. *56* (1990) 908-912.
83. Howie, W.J., and T.V. Suslow. Role of antibiotic biosynthesis in the inhibition of *Pythium ultimum* in the cotton spermosphere and rhizosphere by *Pseudomonas fluorescens*. Mol. Plant-Microbe Inter. *4* (1991) 393-399.
84. Kraus, J., and J.E. Loper. Lack of evidence for a role of antifungal metabolite production by *Pseudomonas fluorescens* Pf-5 in biological control of Pythium damping-off of cucumber. Phytopathology *82* (1992) 264-271.
85. Cuppels, D.A., C.R. Howell, R.D. Stipanovic, A. Stoessl, and J.B. Stothers. Biosynthesis of pyoluteorin: A mixed polyketide-tricarboxylic acid cycle origin demonstrated by [1,2-$^{13}C_2$]acetate incorporation. Z. Naturforsch. *41c* (1986) 532-536.
86. Parkinson, J.S., and E.C. Kofoid. Communication modules in bacterial signaling proteins. Annu. Rev. Genet. *26* (1992) 71-112.
87. Zimmermann, A., C. Reimmann, M. Galimand, and D. Haas. Anaerobic growth and cyanide synthesis of *Pseudomonas aeruginosa* depend on *anr*, a regulatory gene homologous with *fnr* of *Escherichia coli*. Mol. Microbiol. *5* (1991) 1483-1490.
88. Knowles, C.J., and A.W. Bunch. Microbial cyanide metabolism. Adv. Microb. Physiol. *27* (1986) 73-111.
89. Bencini, D.A., C.R. Howell, and J.R. Wild. Production of phenolic metabolites by a soil pseudomonad. Soil Biol. Biochem. *15* (1983) 491-492.
90. Slininger, P.J., and M.A. Jackson. Nutritional factors regulating growth and accumulation of phenazine 1-carboxylic acid by *Pseudomonas fluorescens* 2-79. Appl. Microbiol. Biotechnol. *37* (1992) 388-392.
91. Dowling, D.N., B. Boesten, D.J. O'Sullivan, P. Stephens, J. Morris, and F. O'Gara. Genetically modified plant-microbe interacting strains for potential release into the rhizosphere. In: E. Galli, S. Silver, and B. Witholt (eds.), *Pseudomonas*: Molecular biology and biotechnology, American Society for Microbiology, Washington D.C. 1992, p. 408-414.
92. Sundheim, L., A.R. Poplawsky, and A.H. Ellingboe. Molecular cloning of two chitinase genes from *Serratia marcescens* and their expression in *Pseudomonas* species. Physiol. Mol. Plant Pathol. *33* (1988) 483-491.
93. Van Peer, R., G.J. Niemann, and B. Schippers. Induced resistance and phytoalexin accumulation in biological control of Fusarium wilt of carnation by *Pseudomonas* sp. strain WCS417r. Phytopathology *81* (1991) 728-734.
94. Van Peer, R., and B. Schippers. Lipopolysaccharides of plant-growth promoting *Pseudomonas* sp. strain WCS417r induce resistance in carnation to Fusarium wilt. Neth. J. Pl. Path. *98* (1992) 129-139.
95. Wei, G., J.W. Kloepper, and A. Tuzun. Induction of systemic resistance of cucumber to *Colletotrichum orbiculare* by selected strains of plant growth-promoting rhizobacteria. Phytopathology *81* (1991) 1508-1512.
96. Alström, S. Induction of disease resistance in common bean susceptible to halo blight bacterial pathogen after seed bacterization with rhizosphere pseudomonads. J. Gen. Appl. Microbiol. *37* (1991) 495-501.
97. Zdor, R. E., and A.J. Anderson. Influence of root colonizing bacteria on the defense responses of bean. Plant Soil *140* (1992) 99-107.
98. Naiki, T., and R.J. Cook. Factors in loss of pathogenicity in *Gaeumannomyces graminis* var. *tritici*. Phytopathology *73* (1983) 1652-1656.
99. Simonet, J.M., D. Schneider, J.N. Volff, A. Dary, and B. Decaris. Genetic instability in *Streptomyces ambofaciens*: inducibility and associated genome plasticity. Gene *115* (1992) 49-54.
100. Birch, A., A. Häusler, and R. Hütter. Genome rearrangement and genetic instability in *Streptomyces* spp. J. Bacteriol. *172* (1990) 4138-4142.

101. Hengge-Aronis, R. Survival of hunger and stress: The role of *rpoS* in early stationary phase of gene regulation in *E. coli.* Cell *72* (1993) 165-168.
102. Olsén, A., A. Arnqvist, M. Hammar, S. Sukupolvi, and S. Normark. The RpoS factor relieves H-NS-mediated transcriptional repression of *csgA*, the subunit gene of fibronectin-binding curli in *Escherichia coli.* Mol. Microbiol. *7* (1993) 523-236.
103. Bartlett, D.H., M.E. Wright, and M. Silverman. Variable expression of extracellular polysaccharide in the marine bacterium *Pseudomonas atlantica* is controlled by genome rearrangement. Proc. Natl. Acad. Sci. USA *85* (1988) 3923-3927.
104. Holloway, B.W. *Pseudomonas* in the late twentieth century. In: E. Galli, S. Silver, B. Witholt (eds.), *Pseudomonas.* Molecular biology and biotechnology, American Society for Microbiology, Washington, D.C., 1992, p. 1-8.
105. Oberhänsli, T., and G. Défago. Spontaneous loss of tryptophan side chain oxidase of *Pseudomonas fluorescens* strain CHA0 – a marker for genetic instability. In: C. Keel, B. Koller, and G. Défago (eds.), Plant growth-promoting rhizobacteria – Progress and prospects. IOBC/WPRS Bulletin 14. 1991, p. 392-398.
106. Champness, W., P. Riggle, T. Adamidis, and P. Vandervere. Identification of *Streptomyces coelicolor* genes involved in regulation of antibiotic synthesis. Gene *115* (1992) 55-60.
107. Smit, E., J.D. van Elsas, and J.A. van Veen. Risks associated with the application of genetically modified microorganisms in terrestrial ecosystems. FEMS Microbiol. Rev. *88* (1992) 263-278.
108. Atlas, R.M., G. Sayler, R.S. Burlage, and A.K. Bej. Molecular approaches for environmental monitoring of microorganisms. BioTechniques *12* (1992) 706-717.
109. Olson, B. H. Tracking and using genes in the environment. Environ. Sci. Technol. *25* (1991) 604-611.
110. Gould, W. D., C. Hagedorn, T.R. Bardinelli, and R.M. Zablotowicz. New selective media for enumeration and recovery of fluorescent pseudomonads from various habitats. Appl. Environ. Microbiol. *49* (1985) 28-32.
111. Byrd, J.J., H. Xu, and R.R. Colwell. Viable but nonculturable bacteria in drinking water. Appl. Environ. Microbiol. *57* (1991) 875-878.
112. Islam, M. S., M.K. Hasan, M.A. Miah, G.C. Sur, A. Felsenstein, M. Venkatesan, R.B. Sack, and M.J. Albert. Use of polymerase chain reaction and fluorescent antibody methods for detecting viable but nonculturable *Shigella dysenteriae* type 1 in laboratory microcosms. Appl. Environ. Microbiol. *59* (1993) 536-540.
113. Bentjen, S. A., J.K. Fredrickson, P. Van Voris, and S.W. Li. Intact soil-core microcosm for evaluating the fate and ecological impact of the release of genetically engineered microorganisms. Appl. Environ. Microbiol. *55* (1989) 189-202.
114. Frederickson, J. K., S.A. Bentjen, H. Bolton, S.W. Li, and P. Van Voris. Fate of Tn*5* mutants of root growth-inhibiting *Pseudomonas* sp. in intact soil-core microcosms. Can. J. Microbiol. *35* (1989) 867-873.
115. Kemp, J.S., E. Paterson, S.M. Gammack, M. S. Cresser, and K. Killham. Leaching of genetically modified *Pseudomonas fluorescens* through organic soils: Influence of temperature, soil pH, and roots. Biol. Fertil. Soils *13* (1992) 218-224.
116. Van Elsas, J. D., J.T. Trevors, and L.S. Van Overbeek. Influence of soil properties on the vertical movement of genetically marked *Pseudomonas fluorescens* through large soil microcosms. Biol. Fertil. Soils *10* (1991) 249-255.
117. Annual Book of ASTM Standards. Standard guide for conducting a terrestrial soil-core microcosm test. Designation: E 1197, American Society for Testing and Materials, Philadelphia. Vol. *11.04.* (1988) p. 743-755.
118. Kloepper, J.W., R. Lifshitz, and R.M. Zablotowicz. Free-living bacterial inocula for enhancing crop productivity. Trends Biotechnol. *7* (1989) 39-44.
119. Wüthrich, B., and G. Défago. Suppression of wheat take-all and black root rot of tobacco by *Pseudomonas fluorescens* strain CHA0, isolated from a *Thielaviopsis*-suppressive soil: results of field and pot experiments. In: C. Keel, B. Koller and G. Défago (eds.), Plant growth-promoting rhizobacteria – Progress and prospects. IOBC/WPRS Bulletin 14, 1991, p. 17-22.
120. Loper, J.E., and C.A. Ishimaru. Factors influencing siderophore-mediated biocontrol activity of rhizosphere *Pseudomonas* spp. In: D.L. Keister and P.B. Cregan (eds.), The rhizosphere and plant growth. Kluwer Academic Publishers, Dordrecht, 1991, p. 253-261.

121. Weller, D. M., and R.J. Cook. Suppression of take-all of wheat by seed treatments with fluorescent pseudomonads. Phytopathology *73* (1983) 463–469.
122. Lemanceau, P., P.A.H.M. Bakker, W. J. de Kogel, C. Alabouvette, and B. Schippers. Effect of pseudobactin 385 production by *Pseudomonas putida* WCS385 on suppression of fusarium wilt of carnations by nonpathogenic *Fusarium oxysporum* Fo47. Appl. Environ. Microbiol. *58* (1992) 2978–2982.
123. Fuchs, J., and G. Défago. Protection of tomatoes against *Fusarium oxysporum* f. sp. *lycopersici* by combining an apathogenic *Fusarium* with different bacteria in non-sterilized soil. In: C. Keel, B. Koller and G. Défago (eds.), Plant growth-promoting rhizobacteria – Progress and prospects. IOBC/WPRS Bulletin 14, 1991, p. 51–56.
124. Ownley, B.H., D.M. Weller, and L.S. Thomashow. Influence of in situ and in vitro pH on suppression of *Gaeumannomyces graminis* var. *tritici* by *Pseudomonas fluorescens* 2–79. Phytopathology *82* (1992) 178–184.
125. Paulitz, T.C., and J.E. Loper. Lack of a role for fluorescent siderophore production in the biological control of pythium damping-off of cucumber by a strain of *Pseudomonas putida*. Phytopathology *81* (1991) 930–935.

7 Genetic Strategies to Engineer Expression Systems Responsive to Relevant Environmental Signals

Víctor de Lorenzo

7.1 Introduction

Unlike laboratory bacteria, *G*enetically *E*ngineered *M*icroorganisms (GEMs) destined for deliberate release for environmental and agricultural applications, should express their recombinant genes under the control of signals present in the location where the bacteria are expected to operate (1, 2). The type of signals vary, depending on the specific situation and in some cases they might not be completely defined physically or chemically. Just to raise an example, *Pseudomonas* strains designed to biodegrade *p*oly*c*hloro*b*iphenyls (PCBs) in the rhizosphere of certain plants (3) might be desired to activate transcription of the cognate catabolic pathway only in the vicinity of root exudates; the composition of which might not be well known. Similarly, conditional suicide genetic circuits (4) might be desired which activate lethal genes to avoid proliferation of the GEMs during a certain season of the year (i. e., in a temperature-dependent fashion). Although construction of efficacious expression systems for strains performing in such different conditions can be addressed on a case-by-case basis, the potential demand of specific-purpose recombinant microorganisms in the future has encouraged the development of new broad-host-range expression devices specifically tailored for bacteria in uncontained applications. In this report, I summarize the approaches that we are currently applying to address the problem of expression in the field and how they can be used to solve specific problems in the design of environmental GEMs, in particular, *Pseudomonas, Alcaligenes* and related Gram-negative bacteria.

7.2 Mini-Transposons as Genetic Tools

Most expression systems that we have developed utilize transposon-vectors (5,6) as the basic unit to assemble additional genetic traits. Besides the many advantages of transposons for the construction of GEMs (see below), they permit the engineering of one or more recombinant phenotypes in otherwise wild-type bacteria with a minimal number of manipulations in the laboratory. This may help avoid the loss of competitiveness frequently observed in microorganisms maintained under laboratory conditions for long periods of time.

7.2.1 Rationale for the Utilization of Mini-Tn*5* Transposons

Transposons are among the most useful tools available to the microbial geneticist. The spectrum of different types of transposons, their mechanisms of transposition and their applications in genetic engineering, have been the subject of recent reviews (7). We will discuss here, only those properties of the Tn*5* transposon which are important to the content of this article. Tn*5* is a composite transposon, i. e., its mobility is determined by two insertion sequences (IS*50*) flanking the DNA region determining the selectable phenotype. Tn*5* can be transferred among replicons as a consequence of the action of the transposase, Tnp, encoded by IS*50*R, on cognate short target sequences located at the end of the transposon. While the *tnp* gene is a component of the naturally-occuring Tn*5*, its product still works when the gene is artificially placed outside of the mobile unit, though preferably placed in *cis* to the cognate terminal sequences (7). This allows the construction of recombinant transposons in which only those elements essential for transposition (i. e. IS terminal sequences and transposase gene) have been retained and arranged such that the transposase gene is adjacent to but outside of the mobile DNA segment. Since the elimination of non-essential sequences leads to a major reduction of transposon size, the resulting recombinant mini-Tn*5* transposons are much more convenient to handle than natural transposons.

Two properties of mini-Tn*5* derivatives are particularly useful for applications involving environmental release. First, once inserted in a target sequence mini-transposons are stably inherited and do not provoke DNA re-arrangements or other forms of genetic instability. This is due to the loss of the cognate transposase gene and the greater parts of the IS elements present in the wild-type transposon. Even if host cells later acquire a natural transposon of the same type, Tn*5* transposase work poorly in *trans* (7) and will not stimulate re-transposition of the insertion. A second property is that, due to the loss of the transposase gene during transfer to a new replicon, the host cell does not become immune to further rounds of transposition. This permits the organism to be re-inserted with the same system, provided that subsequent transposons contain distinct selection markers (5).

7.2.2 A Universal Suicide Delivery System

A critical feature of the mini-transposons is the system to deliver them into the target strain, such that its selectable phenotype is expressed only through integration of the transposon into a replicon of the new host. To broaden the range of Gram-negative bacteria which can be targeted with our mini-Tn*5* derivatives, we developed a general suicide delivery system based on the narrow-host range plasmid R6K. Plasmids having the R6K origin of replication require the R6K-specified replication protein π and can be maintained only in host strains producing this protein. Plasmid pGP704 (8) has, in addition to an R6K origin of replication, RP4*oriT*, the origin of transfer sequence of the conjugative plasmid RP4. pGP704 is maintained stably in λ*pir* lysogens and can be mobilised into target bacteria through RP4 transfer functions. Bacteria receiving pGP704 but lacking the π protein are generally non-permissive and do not maintain the transferred plasmid. Since mini-Tn*5*s and their cognate transposase genes are constructed as inserts in pGP704 (Fig.1), selection of exconjugants stably expressing the marker of the mini-transposon selects for clones in which they have been transferred to a replicon of the recipient. Transposition is promoted by the cognate transposase encoded on the same plasmid proximal but external to the transposon.

The basic structure the mini-Tn*5* transposons and their delivery system (Fig.1) satisfies a number of requirements to engineer very efficient expression systems, as discussed below. The basic experimental design for their use is identical in all cases regardless of the specific application or the target strain. These transposons have over the last few years found broad utilization for manipulations of a variety of bacteria including *E. coli, Klebsiella, Salmonella, Proteus, Vibrio, Bordetella, Actinobacillus, Rhizobium, Rhodobacter, Agrobacterium, Alcaligenes* and several pseudomonads.

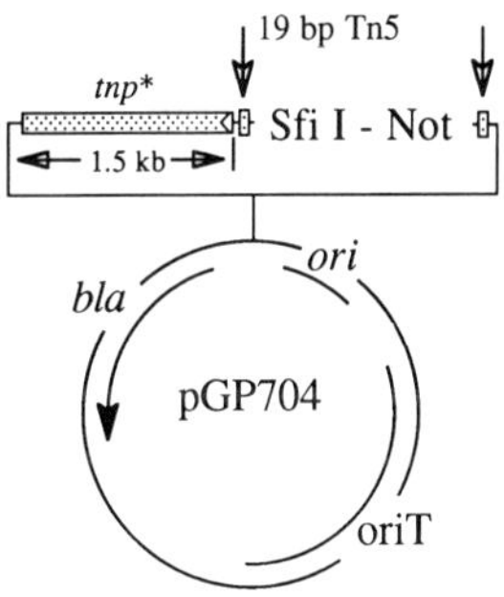

Fig. 1. Basic Organization of Tn*5*-based Transposon Vectors and their Suicide Delivery Plasmid. The Figure represents the features common to all constructions, the whole arrangement being generally named as the pUT system. The transposase gene of Tn*5* devoid of *Not* I sites (*tnp**, 1.5 kb) has been cloned in the 3.7 kb suicide vector pGP704 (8), nearby but outside a short polylinker which includes sites for two unique and very unusual restriction enzymes, *Sfi*I and *Not*I. This sequence is flanked by the two 19 bp I and O termini of Tn*5*. The *Sfi*I site is then used for the insertion of a variety of antibiotic and non-antibiotic selection markers whereas the *Not* I site is used for cloning DNA fragments for insertion into the chromosome of target bacteria. These constructions can be maintained only in donor host strains that produce the R6K-specified π protein, which is an essential replication protein for R6K plasmid and derivatives. Donor plasmids containing hybrid transposons are transformed into a λ*pir* -lysogen *E. coli* strain with a chromosomally integrated RP4 that provides conjugal transfer functions. Delivery of the donor plasmids into target bacteria is accomplished through a simple mating protocol (6), followed by selection of the mating mixture for transposon markers and counter-selection of the donor strain.

7.2.3 Alternative Selection Markers

Although the use of genes encoding resistances to antibiotics of clinical value are at the basis of molecular genetics, their utilization in non-contained applications is perceived to be undesirable in bacteria designed for release in large quantities into the environment. Some alternatives have been developed (Table 1), including the use of resistances to herbicides such as bialaphos (5) and glyphosate (9), nutritional markers like growth on lactose as the only carbon source and resistances to heavy metal ions like mercury (5), arsenite (5) or tellurite (10). Most of these have been combined with mini-Tn5 transposons to make mobile elements devoid of antibiotic resistances. Nutritional markers (capacity to grow on unusual carbon sources) are particularly useful for genera such as *Pseudomonas cepacia*, which quickly develops spontaneous tolerance to most inhibitory agents. A second alternative, not yet used with transposons but successfully attempted with plasmids, is the use of auto-selection markers such as *thyA*$^{+}$, which compensate for the loss or deletion of a gene essential for survival of the target cells (11). A final and most promising approach developed recently to avoid the propagation of antibiotic resistances is the chromosomal insertion of heterologous DNA sequences through homologous recombination of flanking DNA stretches with their chromosomal counterparts (12). This last approach, which was flawed due to the difficulty in selecting double recombination events is now finding widespread use due to the availability of conditionally lethal genes such as *sacB* of *Bacillus,* which once included in the delivery vector permits selection for the loss of the plasmid portion of the system upon exposure of the cells to sucrose (12,13). The result of this genetic device is the chromosomal insertion of the heterologous DNA sequence and the loss of any vector marker.

Tab. 1. Environmentally-safe, Alternative Selection Markers for GEMs destined for Environmental Release

Marker gene	Origin	Selectable phenotype	Refs.
bar	*Streptomyces hygroscopicus*	Resistance to herbicide bialaphos	5
aroA CT7	*Salmonella typhimurium*	Resistance to herbicide glyphosate	9
merTPAB	*Serratia marcescens*	Resistance to mercuric salts/ organomercurials	5
arsAB	*E. coli* /R773	Resistance to arsenite	5
lacZY	*E. coli*	Growth on lactose	2
thyA	*Lactococcus lactis*	Auto-selective in *thyA*$^{-}$ strains	11
tel	RP4 plasmid	Resistance to potassium tellurite	10

7.3 Engineering Gene Expression within Mini-Transposons

Gene expression is the result of transcriptional and post-transcriptional events. In spite of the many broad host-range systems available (14), expression of a recombinant phenotype may be unpredictable when the time comes to express an heterologous activity in microorganisms poorly characterized genetically, as is the case in many Gram-negatives of environmental interest. Transposon-vectors allow, however, to probe and to utilise indigenous promoters for expression of recombinant genes even in the absence of any information on promoter structure or regulation in the native host.

7.3.1 Selecting an Adequate Level of Transcription

The most straightforward procedure to have a gene or operon transcribed in response to a pre-determined signal is to construct a specialized mini-Tn*5* transposon in which the promoterless gene or gene cluster is placed next to the I or O termini of the mobile element (Fig. 2). The strain of interest is then mutagenized with such a transposon, and insertions which happen to have occurred next to a promoter regulated by that signal are screened, for instance, with immunological procedures. Although this approach (random probing of a host promoter as a source of transcription activity) has not been used yet in environmental applications, the concept has been successfully employed to engineer expression of *Pertussis* toxin in *bvg*-negative *Bordetella bronchiseptica* (15) for live-vaccine development.

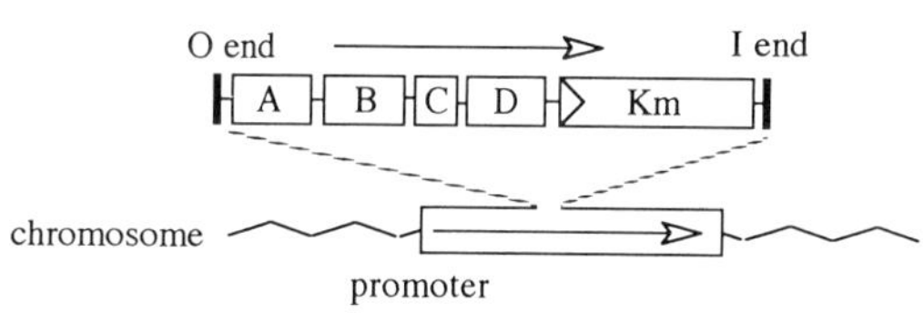

Fig. 2. Random Probing of Specific Levels of Expression. This approach is particularly interesting in cases where information on promoters functional in the target strain is scarce. A hybrid transposon is constructed on the basic structure of the pUT system (see Fig.1) so that the 5′ extreme end of the promoterless gene or gene cluster to be expressed in the target bacterium is placed nearby the O or I ends of the mobile element. The resulting construction is then used to generate random insertions of the element throughout the chromosome. Insertions are selected through the marker present within the mini-transposon (resistance to kanamycin in this case). As sketched in the figure, expression of the gene cluster ABCD may become regulated by the promoter which happens to occur upstream of the insertion, the strength (and sometimes the regulation) of which can be predetermined. A good choice is to examine constitutive expression of the recombinant gene at an adequate level through, for instance, immunodetection (15) of the corresponding gene product(s).

In many cases, however, construction of specialized transposons of this sort is impractical and some information on promoter strength and its regulation might be required prior to engineering expression of an heterologous gene placed artificially downstream. The range of promoter strengths available in a given Gram-negative strain can be assessed with mini-Tn*5* derivatives carrying a variety of convenient reporter genes such as *lacZ, xylE* or *luxAB* (6). The resulting insertions not only provide a repertoire of promoter strengths, as revealed by the activities of the corresponding reporter products, but they also offer a genetic "hook" to place expression of heterologous genes under the same regulation of the reporter gene through homologous recombination of specialized plasmid vectors with transposon sequences inserted into the chromosome of target bacteria (de Lorenzo, unpublished).

7.3.2 Post-Transcriptional Bottlenecks

While random generation of gene fusions to reporter genes permit detection of transcriptional activity in many different bacterial species, transcription by itself may not be the limiting step for heterologous gene expression. Translation initiation regions (TIRs) behave differently in disparate hosts and the performance of a given TIR in a poorly characterized target bacterium is somewhat unpredictable. One promising possibility to overcome the problem is the use of translational enhancers (16) to ensure translation of a particular sequence in a wide variety of hosts. These are short regions (40–50 base pairs) found typically in plant viruses which can functionally substitute the Shine-Delgarno sequence which typically precedes the first codon of structural sequences in bacteria. A translational enhancer of the tobacco mosaic virus permitted expression of a *cat* gene (Chloramphenicol acetyl transferase) (in *E. coli, S. typhimurium, E. amylovora, A. tumefaciens, A. rhizogenes, R. meliloti and X. campestris* (16) when placed at different distances (19, 35 or 110 base pairs) upstream of the initiator AUG codon. Adaptation of these sequences to transposon-based expression devices is currently in progress.

An additional bottleneck which may hinder expression of recombinant genes in an heterologous Gram-negative host is mRNA stability. Although the gross determinants of mRNA half-lives are probably shared by virtually all bacteria, the effect of secondary structures or nucleotide sequences on stability of specific mRNAs in particular hosts might again be unpredictable. An interesting approach to solve this problem has been developed based on the properties of the bacteriophage T4 gene 32. The gene 32 mRNA has the unusual feature of being extremely stable due to the structure and processing of its leader untranslated 5-end (17). Hybrid constructions consisting of the gene 32 native promoter/TIR region and *xylE,* expressed the reporter product at high levels in a whole variety of Gram-negative bacteria, including *Agrobacterium tumefaciens* and *Xanthomonas maltophilia.* This suggests that the structural determinants at the 5′ end of the gene 32 mRNA which increase stability of the transcript in *E. coli,* may be functional also when fused to heterologous genes and placed in heterologous hosts (17).

7.4 Engineering Alkyl- and Halo-aromatic Responsive Phenotypes

Since aromatic compounds are a major source of environmental pollution, it is frequently desirable to have some phenotypes expressed as a response to their presence. For this purpose, a number of recombinant mini-Tn*5* transposons (Fig.3) were constructed which contain outward-facing *Pm, Pu* or *Psal* promoters from the catabolic plasmids TOL (Toluene degradation) and NAH (Napthalene degradation) of *Pseudomonas putida,* along with their cognate wild type regulatory genes (*xylS, xylR, nahR*) or mutant varieties (*xylS2*). Transcription from such promoters is activated when host bacteria face certain aromatic compounds, such as alkyl- and halobenzoates (XylS, XylS2), alkyl- and halo/toluenes (XylR) or salicylates (NahR). These transposons enable the generation of conditional mutations dependent on the presence of specific effectors, as well as the engineering of strains expressing heterologous genes that are regulated by aromatic inducers (18). The broad host-range of the Tn*5* transposition system and the stability of the inserted genes make this system suitable for the construction of stable strains exhibiting halo/alkyl aromatic-regulated conditional phenotypes in the absence of antibiotic selection. This may be a regulatory requirement for some uncontained bioremediation and biomonitoring applications.

In our hands, the most practical construction for *Pseudomonas putida* was that derived from *xylS/Pm*, since it had a relatively low basal level of expression, which is quickly triggered (in monocopy gene dosage) to an equivalent strength of 10–14 Miller units of ß-galactosidase upon addition of the inducer *m*-toluate (18). The induction pattern of mini-Tn*5 xylR/Pu* was particularly interesting, because *Pu* activity is virtually indistinguishable from background in the absence of inducer, whereas it reaches maximum activity after overnight exposure to vapours of the effector *m*-xylene. Transcriptional activity of *Pu* is subjected to various controls which include not only XylR in combination with the σ^{54}-containing RNA polymerase, but also integration host factor (19) and growth rate regulation. The result of this is the absence of significant transcription in uninduced cells and low or no activity of the *Pu* promoter during the exponential growth of induced cells. This property could in principle be exploited for cloned genes whose products are deleterious for growing cells. Furthermore, since the expression system can be activated upon exposure to vapours of inducer rather than direct addition to the medium, activation of a desired gene may be arranged in a non-disruptive manner. These transposons containing conditional catabolic promoters have been combined also with the T7 polymerase gene to engineer expression of multiple genes in response to single aromatic compounds (20).

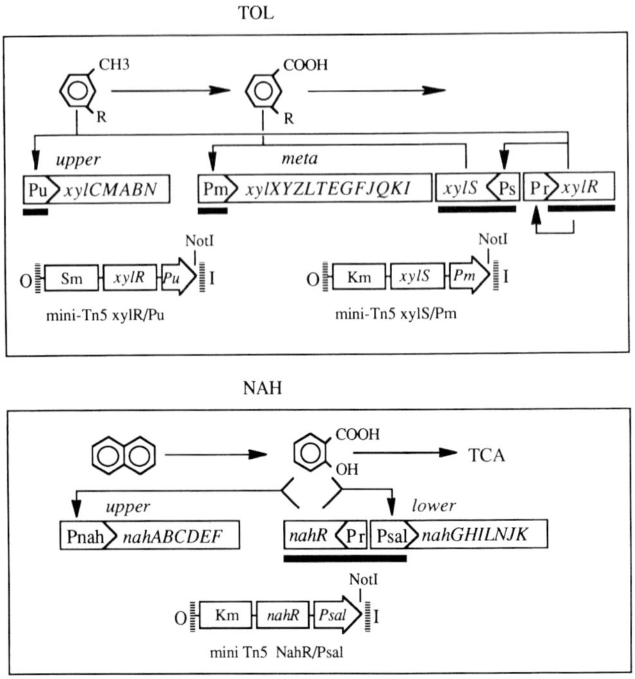

Fig. 3. Expression Systems Based on Catabolic Promoters of *Pseudomonas.* **The upper box** (TOL system) shows the salient features of the regulation of the TOL plasmid (biodegradation of toluene and *m-/p*-xylenes) of *Pseudomonas putida,* which were used to develop transposon-based expression system containing conditional catabolic promoters as the source of transcriptional activity (18). The scheme (not to scale) summarizes the regulatory cascade which controls transcription of TOL genes. In the presence of pathway substrates like *m*-xylene, *Pu* (promoter of the *upper* operon) and *Ps* (promoter of *xylS* gene) are activated by XylR in combination with the σ^{54}-containing form of RNA polymerase. Subsequently, an excess of XylS product or XylS bound to its effectors (i.e., substrates of the *meta*-pathway like 3-methyl benzoate) activates *Pm.* Finally, the XylR product seems to auto-regulate its own transcription (represented as a solid arrow). The result of TOL activities is the mineralization of the aromatic substrates down to TCA cycle intermediates. The regulatory elements of the pathway exploited for construction of the expression systems are underlined with a solid bar. Below the scheme of the pathway, the organization of the hybrid-transposons mini-Tn*5 xylR/Pu* and mini-Tn5 *xylS/Pm* are shown. These transposons contain respectively the *Pu* and *Pm* catabolic promoters pointing towards a *Not*I site, along with their cognate regulatory genes *xylR* or *xylS.* An additional transposon, mini-Tn5 *xylS2/Pm* was constructed also (not shown) which includes a mutant variety of *xylS* responsive to a wider variety of benzoate derivatives. For selection purposes, a resistance marker (streptomycin, Sm or kanamycin, Km) has been included also within the mobile unit. Expression of heterologous genes cloned at the *Not*I site of the transposon is induced upon exposure of the cells inserted with the hybrid construction to TOL inducers.

The lower box (NAH system) shows the regulatory features of the naphtalene degrading pathway determined by the NAH7 plasmid of *P. putida.* Upon exposure of the cells to naphtalene vapours, the basal expression of the *upper* operon is enough to generate a certain amount of intracellular salicylate. This intermediate binds the product of the regulatory gene *nahR,* which in turn further activates the *Pnah* promoter of the *upper* operon and the *Psal* divergent promoter of the *lower* pathway. Unlike the TOL system, a single inducer (salicylate) activates the *upper* and *lower* catabolic routes. NahR belongs to the LysR family of transcriptional regulators. The regulatory elements of the pathway included in the mini-Tn*5 nahR/Psal* transposon represented below the scheme of the pathway, are underlined. The gross organization of this mobile element is identical to that of the counterparts of the TOL pathway, and it permits one to engineer expression of heterologous genes in response to salicylate.

7.5 Outlook

The simple organization of mini-transposons has made them a sort of "Swiss army knife" to construct GEMs destined for environmental applications. The possibility now exists not just to release into the environment predictably modified laboratory strains, but to improve the phenotype of naturally occurring bacteria by inserting in their chromosome one or more hybrid transposons through a very simple mating protocol (6). There are, however, a few areas within the subject of vector development which require further attention because they become major limitations in very significant cases. One specific feature to be improved is the transposon delivery system. Poor yield (or no yield at all) of exconjugants reflects in most instances low frequency of RP4-mediated transfer. Several properties of target cells, such as restriction systems, surface exclusion phenomena or lack of essential envelope functions, may lead to low or zero transfer. If the target strain is unable to act as a recipient for RP4, then most of the constructions described in this article are of little use, because the whole transposon delivery system is based on having an efficient conjugation. Alternative strategies for suicide delivery of transposons are under way which include broad host-range conditional replicons capable of becoming established in the target strain but can then be subsequently eliminated by stopping expression of essential replication proteins (de Lorenzo & Díaz, unpublished).

Another foreseeable area of vector development is the utilization of starvation promoters for heterologous expression purposes in the field. Since growth of soil bacteria occurs under nutrient conditions unable to support exponential growth, nutrient starvation is considered a universal signal, potentially useful to express heterologous genes in the environment. Promoters responsive to carbon, nitrogen, iron and phosphate starvation have been characterized in many Gram-negative bacteria and they are in principle excellent assets as building blocks for expression systems (21). At least in *E. coli,* certain subsets of proteins examined in two-dimensional gels are expressed in response to depletion of specific nutrients (22), indicating the presence of a variety of promoters ultimately responsible for the effect. An alternative sigma factor, σ^S (the product of *rpoS*) seems to be directly or indirectly involved in expression during late growth stages and stationary phase (23), but other factors such as the proteins H-NS (24), UspA (22) and Dps (25) influence or determine gene expression when cells have stopped growing. To facilitate the identification of promoters particularly active under starvation conditions in a variety of Gram-negative bacteria, we have used specialized transposons as genetic probes to detect promoters of genes which are preferentially expressed during late stages of growth in a variety of Gram-negative bacteria. One such transposon generates random fusions to an artificial bi-cistronic operon composed of a promoterless *lacZ* gene coupled to a *tet* cassette, which confers a tetracycline-resistant phenotype when co-transcribed with the leading reporter gene. Random insertion in front of growth-phase dependent promoters give rise to exconjugant colonies with distinct phenotypes (26). Further development of the system for heterologous expression purposes include the construction of a plasmid capable of delivering, through homologous recombination, a promoterless gene (or genes) in front of the starvation promoter, and its combination with the T7 polymerase gene.

Acknowledgments

This work was supported by grants BIO89-0497 and BIO92-1018-CO201 of the Spanish Comisión Interministerial de Ciencia y Tecnología (CICYT).

7.6 References

1. de Lorenzo, V. Genetic engineering strategies for environmental applications. Current Opinion in Biotechnology *3* (1992) 227-231.
2. de Lorenzo, V. and K.N. Timmis. Specialized Host-Vector systems for the engineering of *Pseudomonas* strains destined for environmental release. In: *Pseudomonas*: Molecular Biology and Biotechnology. E. Galli, S. Silver & B. Witholt (Eds.). American Society for Microbiology, Washington, 1992, p. 415-428.
3. Walton, B.T. and T.A. Anderson. Plant-microbe treatment systems for toxic waste. Curr. Opinion Biotech. *3* (1992) 267-270.
4. Molin, S., P. Klemm, L.K. Poulsen, H. Biehl, K. Gerdes and P. Andersson. Conditional suicide system for containment of bacteria and plasmids. Bio/Technology *5* (1987) 1315-1318.
5. Herrero, M., V. de Lorenzo and K.N. Timmis. Transposon vectors containing non-antibiotic selection markers for cloning and stable chromosomal insertion of foreign DNA in Gram-negative bacteria. J. Bacteriol. *172* (1990) 6557-6567.
6. de Lorenzo, V., M. Herrero, U. Jacubzik and K.N. Timmis. Mini-Tn*5* transposon derivatives for insertion mutagenesis, promoter probing and chromosomal insertion of cloned DNA in Gram-negative eubacteria. J. Bacteriol. *172* (1990) 6568-6572.
7. Berg, C.M., D.E. Berg and E.A. Groisman. Transposable elements and the genetic engineering of bacteria. In: Mobile DNA. D.E. Berg and M.M. Howe (Eds.), American Society for Microbiology, Washington, D. C., 1989, pp. 879-926.
8. Miller, V.L. and J. Mekalanos. A novel suicide vector and its use in the construction of insertion mutations: osmoregulation of outer membrane proteins and virulence determinants in *Vibrio cholerae* requires *toxR*. J. Bacteriol. *170* (1988) 2575-2583.
9. Stalker, D.M., W. Hiatt and L. Comai. A single amino acid substitution in the enzyme 5-enolpyruvylshikimate-3-phosphate synthase confers resistance to the herbicide glyphosate. J. Biol. Chem. *260* (1985) 4724-4728.
10. Walter, E. and D. Taylor. Comparison of tellurite resistance determinants from the IncPa plasmid RP4Te^r and the IncHII plasmid pHH1508a. J. Bacteriol. *171* (1989) 2160-2165.
11. Ross, P., F. OGara and S. Condon. Thymidylate synthase gene from *Lactococcus lactis* as a genetic marker: An alternative to antibiotic resistance genes. Appl. Environ. Microbiol. *56* (1990) 2164-2169.
12. Kaniga, K. and G. Cornelis. A wide-host range suicide vector for improving reverse genetics in Gram-negative bacteria: inactivation of the *blaA* gene of *Yersinia enterocolitica*. Gene *109* (1991) 137-141.
13. Kamoun, S., E. Tola, H. Kamdar and C. Kado. Rapid generation of directed and unmarked deletions in *Xanthomonas*. Mol. Microbiol. *6* (1992) 809-816.
14. Schmidhauser, T.J., G. Ditta and D.R. Helinski. Broad host range plasmid cloning vectors for Gram-negative bacteria. In: Vectors: A survey of molecular cloning vectors and their uses. R.L. Rodriguez and D.T. Denhardt (Eds.), Butterworths. Boston, 1987, pp. 287-332.
15. Walker, M., M. Rohde, J. Wehland and K.N. Timmis. Construction of mini-transposons for constitutive and inducible expression of *Pertussis* toxin in *bgv*-negative *Bordetella bronchiseptica*. Infect. Imm. *59* (1991) 4238-4248.
16. Gallie, D. and C. Kado. A translational enhancer derived from tobacco mosaic virus is functionally equivalent to a Shine-Dalgarno sequence. Proc. Natl. Acad. Sci. USA *86* (1989) 129-132.

17. Frey, J., E.A. Mudd and H. Krisch. A bacteriophage T4 expression cassette that functions efficiently in a wide range of Gram-negative bacteria. Gene *62* (1988) 237–247.
18. de Lorenzo, V., S. Fernández, M. Herrero, U. Jakubzik and K.N. Timmis. Engineering of alkyl- and haloaromatic gene expression with mini-transposons containing regulated promoters of biodegradative pathways of *Pseudomonas*. Gene *130* (1993) 41–46.
19. de Lorenzo, V., M. Herrero, M. Metzke and K.N. Timmis. An upstream XylR- and IHF-induced nucleoprotein complex regulates the s^{54}-dependent *Pu* promoter of TOL plasmid. EMBO J. *10* (1991) 1159–1167.
20. Herrero, M., V. de Lorenzo, B. Enslcy and K. Timmis. A T7 RNA polymerase-based system for the construction of *Pseudomonas* strains with phenotypes dependent on TOL,-*meta* pathway effectors. Gene *134* (1993) 103–106.
21. Matin, A. The molecular basis of carbon-starvation-induced general resistance in *Escherichia coli* . Mol. Microbiol. *5* (1991) 3–10.
22. Nyström, T. and F. Neidhardt. Cloning, mapping and nucleotide sequencing of a gene encoding a universal stress protein in *Escherichia coli*. Mol. Microbiol. *6* (1992) 3187–3198.
23. Lange, R. and Hengge-Aronis. Identification of a central regulator of stationary-phase gene expression in *Escherichia coli*. Mol. Microbiol. *5* (1991) 49–59.
24. Hulton, C., A. Seirafi, J. Hinton, J. Sidebotham, L. Waddel, G. Pavitt, T. Owen-Hughes, A. Spassky, H. Buc and C.F. Higgins. Histone-like protein H1 (H-NS), DNA supercoiling and gene expression in bacteria. Cell *63* (1990) 631–642.
25. Almirón, M., A. Link, D. Furlong and R. Kolter. A novel DNA binding protein with regulatory and protective roles in starved *Escherichia coli*. Genes Develop. *6* (1992) 2646–2654.
26. de Lorenzo, V., I. Cases, M. Herrero and K. N. Timmis. Early and late responses of TOL promoters to pathway inducers: Idendification of postexponential promoters in *Pseudomonas putida* with *lacZ-tet* bicistronic reporters. J. Bacteriol. *175* (1993) 6902–6907.

8 Genetic Typing of Microorganisms: Current Concepts and Future Prospects

Hans-Volker Tichy and Reinhard Simon

8.1 Introduction

In areas like epidemiology and ecology, as well as biological safety, it is important to identify species and strains of organisms accurately. Typically, identification and typing of bacterial isolates is done by characterizing phenotypical properties. Utilization of different substrates, presence or absence of certain enzymes, immunological cross-reactivity, phage or antibiotic sensitivity and isoenzyme comparisons are commonly used to classify strains.

More recently, methods based on genome analysis of organisms are gaining increasing attention. These methods are not influenced by varying growth conditions or culture age which may alter the synthesis of enzymes, metabolites or surface components or their accessibility for detection purposes.

In eucaryotes, the detection of sequence variability allowed the development of genetic "fingerprinting" methods. These polymorphisms are useful as markers for the mapping of genes and as a means to distinguish and compare individuals on a genetic basis. In 1979, it was discovered by Jeffreys (1) that the presence of sequence variations in the human genome lead to differences in the size of DNA fragments produced when the DNA was cut with specific restriction enzymes. These variations result from base exchanges leading to the appearance of new or disappearance of existing restriction sites.

Some marker systems used for genetic mapping show extreme variability between restriction fragments from different individuals. Wyman and White (2) were the first to describe the identification of a highly polymorphic locus in human DNA. These variabilities do not result from single base changes, instead they are caused by the variable number of short (up to 60 bp), tandemly repeated sequences. Jeffreys (3) used these polymorphisms to generate genetic "fingerprints" of human individuals. Nakamura et al. (4) proposed the acronym VNTR (Variable number of tandem repeats) for this type of polymorphism. The variation described above lead to restriction fragment length polymorphisms (RFLP). The detection of these RFLPs originally relied on the digestion of isolated total DNA, gel electrophoresis and Southern hybridizations. Recently, the application of the Polymerase Chain Reaction (PCR) for amplification of the polymorphic regions has simplified the procedure.

In bacteria, VNTRs are not found, but sequence variabilities do occur due to single base changes. Thus, RFLP analyses can be used as a genetic typing method for bacterial strains. A useful class of bacterial RFLP fingerprinting detects sequence variations within the operons coding for ribosomal RNAs which contain highly conserved as well as variable regions. The conserved regions are used to detect DNA restriction fragments containing rRNA genes whereas the variable regions provide the basis for the RFLPs.

A more general approach for genetic typing of strains of any procaryotic or eukaryotic species is the RAPD (Random Amplified Polymorphic DNA) analysis. Short PCR primers of arbitrarily chosen sequence are used to amplify regions of the genome which, after separation of the resulting DNA fragments, produce a band pattern useful as a strain- (or individual-) specific "fingerprint" of the genome. A similar approach using tRNA consensus sequences as PCR-primers generates species-specific fragment patterns for bacteria, which can be used to assign members of the same genus to a species, after which strains can be further analyzed by applying high resolution RAPD fingerprinting.

In this paper, we will describe the different approaches to DNA fingerprinting of bacteria and discuss their advantages and fields of application focusing on the use of the PCR process as a tool for genetic typing.

8.2 Techniques for the Analysis of DNA Sequence Polymorphisms

8.2.1 Southern Blot and Hybridization

To allow detection of a DNA sequence polymorphism, a gene probe complementary to the polymorphic region is needed. The probe is radiolabeled or labeled with a non-radioactive reporter group and allowed to hybridize to the target DNA. To obtain a pattern that can be compared between individuals, the target DNA is digested with one (or several) restriction enzyme(s) and the fragments separated by electrophoresis on an agarose gel.

Using the Southern Blotting method, the DNA fragment pattern is then transferred onto the surface of a solid support. The labeled probe is added and allowed to hybridize to the target DNA fragments. After performing a detection process, hybridization signals revealing the probe/target hybrids appear. Since the fragments were previously separated according to their molecular weight on the gel, hybridizing fragments can be classified and compared by their sizes. If a DNA region shows a polymorphism, samples from different individuals yield hybridizing fragments of different molecular weights.

8.2.2 PCR-Amplification of Polymorphic DNA

The Southern blot hybridization described above is a rather time consuming, multistep process. The advent of the polymerase chain reaction (PCR) technique allowed the development of a faster technique for the analysis of polymorphisms. A PCR results in the amplification of a specific DNA sequence and is performed by an *in vitro* replication system. Starting from a single copy of a gene, theoretically 10^9 copies are present after performing 30 cycles of amplification. Each cycle consists of three steps: (i) denaturation of the target DNA at 94 °C, (ii) annealing of two primers flanking the sequence to be amplified at 55 °C, and (iii) extension of the primers at 72 °C, provided that a thermostable polymerase is used. After each cycle, the number of DNA molecules has theoretically doubled, thus leading to exponential amplification of the target sequence. The use of thermostable DNA polymerases such as *Taq* for the reaction allows the whole process to be performed without adding fresh enzyme after each cycle, allowing automation of the process.

Using primers flanking a polymorphic region, the enclosed fragment can be amplified to a high degree. The products of the PCR reaction itself (in the case of a VNTR-polymorphism) or the fragments resulting from a restriction enzyme digestion (in the case of polymorphisms due to single base changes) can be detected directly on ethidium bromide stained agarose gels. The sizes of the amplified or restriction enzyme cut fragments correspond to the sizes of the hybridizing fragments in the Southern hybridization technique previously described. Thus, the analysis of polymorphic regions can be reduced to only three or four steps (5): Preparation of the sample, amplification, restriction enzyme digestion, if needed, and gel electrophoresis. In contrast, six steps, some of them very time-consuming, are necessary for a Southern analysis (DNA isolation, restriction enzyme digestion, blotting, labeling of the probe, hybridization, and detection).

8.2.3 Ribotyping of Bacterial Strains

In prokaryotic organisms, no hypervariable sequences corresponding to the VNTRs of higher organisms are found. Sequence polymorphisms in bacteria only result from single base changes adding or removing restriction sites. Although theoretically any gene could be used to type bacterial strains by RFLP analyses, this would necessarily involve the isolation of different gene probes for different bacterial species. A more general approach is based on the high conservation of certain essential cell components, mainly the RNA moiety of the ribosomes. A single probe encompassing the DNA coding for the 16S and 23S rRNAs isolated from one bacterial species can be used as a probe for a wide range of other bacterial species. Traditionally, this analysis called "Ribotyping" (6) is done by a Southern blot hybridization experiment using restriction enzyme cut genomic DNA as target and labeled ribosomal RNA or rDNA as a probe, but the conservative nature of the rRNA genes also allowed the design of generally applicable PCR primers for this purpose (7), thus simplifying the procedure analogous to the analysis of VNTR and other polymorphisms by PCR amplification as discussed in the preceeding paragraph.

Using the primers complementary to conserved regions at the beginning of the 16S rRNA gene and the end of the 23S rRNA gene the DNA region between these binding sites is amplified. The fragment obtained is cut with a restriction enzyme and the products analyzed by agarose gel electrophoresis. If the same restriction enzyme is used for digestion of total DNA in the case of ribotyping and for digestion of the amplified fragment in the case of the PCR version, identical fragment patterns are obtained. Only the leftmost and rightmost fragments are different since one end of them is defined by a primer binding site rather than a restriction site in the PCR version of ribotyping. Fragments appear as signals after hybridization of a Southern blot if a traditional RFLP analysis is done or directly, as DNA fragments on an agarose gel.

8.2.4 Fingerprinting by Arbitrarily Primed PCR

It has been observed that by using a single arbitrarily chosen primer for a PCR amplification, discrete and reproducible sets of products can be generated from most bacterial and eukaryotic genomes (8, 9). No prior sequence information is necessary for this method to be applicable. The resolution of the patterns is high enough to allow distinguishing different strains of the same species. This technique was originally called AP-PCR (Arbitrarily primed PCR, 8) or RAPD (Random amplified polymorphic DNA, 9). Compared with other typing methods, PCR-fingerprinting is fast and easy to perform that gives as much or more information than other methods.

As starting material, cells from a single bacterial colony or, in the special case of *Rhizobium*, bacterial cells directly isolated from a single root nodule of a leguminous plant (10) are sufficient to generate a fingerprint. The cells are lysed by alkali, detergent and heat and a small aliquot of this crude preparation added as template DNA for the PCR. Short (down to 10 bp) oligonucleotides of arbitrarily chosen sequence are used as primers. After denaturation of the template DNA, the primer molecules bind to target sequences of full or partial complementarity (Fig 1.). If two target sites are present on different DNA strands within a distance less than a few kb and oriented properly, the DNA region in between is amplified during the PCR. To obtain a sensitivity of detection in the nanogram range, the amplification products are separated on ultrathin polyacrylamide gels and stained with silver (11). The number of species of organisms characterized by RAPD analysis is rapidly increasing and includes bacteria of the genera *Staphylococcus* and *Streptococcus* (8) *Escherichia coli* (12), *Frankia* (13), *Borrelia burgdorferi* (14), and *Rhizobium leguminosarum* (10), fungi like *Fusarium* and *Agaricus* (15, 16), higher plants, eg. alfalfa (17), corn (18), potato (19) banana (20), tomato (21), and soybean (9, 22) and mammals, eg. mouse (23, 24).

Using a set of primers alone or in combination, numerous sets of fingerprint patterns can be generated allowing unequivocal identification of bacterial strains. However, not all primer sequences are equally efficient in priming DNA synthesis from

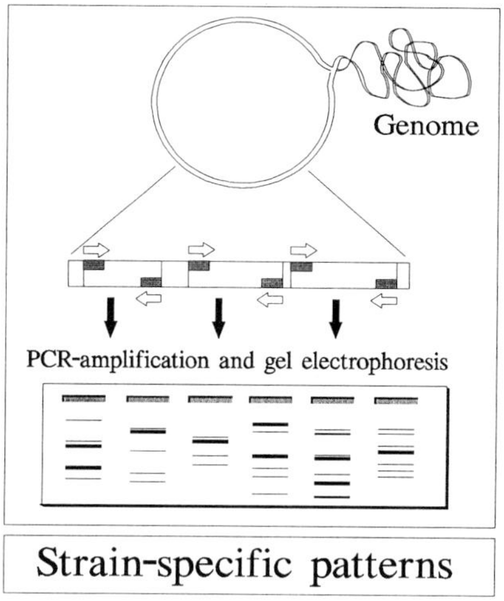

Fig. 1. Schematic Representation of PCR Fingerprinting with Arbitrary Primers. Only a single type of primer is used, for which binding sites (small dark boxes) are present at random sites within a genome. During the Polymerase Chain Reaction process, the PCR-primers (arrows) bind to these sites, amplification of DNA sequences encompassed by them occurs and the resulting DNA fragments are separated and visualized on a polyacrylamide gel. The patterns obtained show differences between strains within a species.

any template. A primer GC content (50% to 100%) higher than the GC-content of the organism under investigation is preferred and several primers have to be tested for reproducibility of the results for each bacterial species. Optimization seems to be especially important for high-GC organisms such as *Rhizobium* (59%-64% GC) or *Micrococcus* (69%-73% GC), the latter being the extreme. Caetano-Anollés et al. (25) introduced the application of the Stoffel fragment of *Taq* polymerase (lacking exonuclease activity) for use in PCR-fingerprinting. In our experience, using a thermostable polymerase lacking exonuclease activity gives a greater number of amplification products especially in these high GC cases, whereas in the case of *E. coli* (50% GC) no further improvement of the already complex patterns obtained with unmodified *Taq* could be observed.

We used PCR-fingerprinting to analyze *R. meliloti* strains (Fig. 2). Since these were were previously typed by RFLP fingerprinting (B. Kosier, R. Simon, unpublished results), we had a measure for the quality of classifications by PCR-fingerprinting. By comparing the groupings of strains obtained with IS-element-fingerprinting (RFLP typing using IS-elements as probes, 26) with the grouping according to RAPD patterns, we found a complete consistence of the results of both approaches. Since RAPD fingerprinting is much faster and less expensive than RFLP analyses, it could replace the latter once our findings and the fingerprinting data of other groups are based upon large scale comparisons, one of which is currently under way in our lab.

8.2.5 Fingerprinting by tRNA Consensus Primed PCR

Application of the RAPD analysis for fingerprinting gives patterns too diverse to let a classification of strains into species become obvious. Thus, it is necessary to group strains exactly into species before applying high-resolution fingerprinting. In some

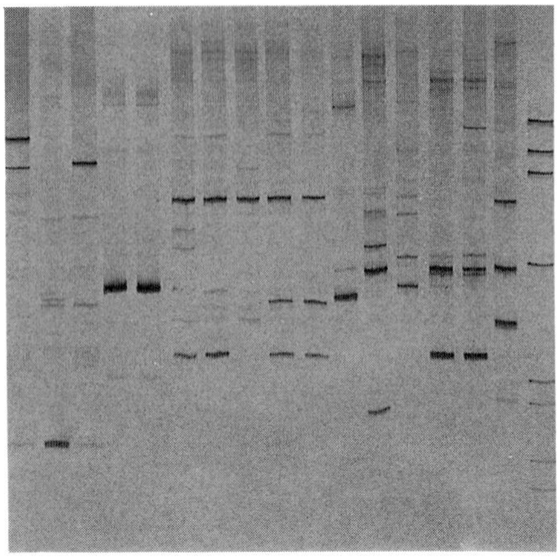

Fig. 2. PCR Fingerprinting of *Rhizobium* Strains. Two 10 bp primers (DAF-3 and DAF-4) were used for amplifications with crude lysates of single colonies as DNA source. Lanes 1 through 5 and lanes 6 through 10 show amplifications done with DAF-3 and DAF-4, respectively. Lanes 1 and 6 show *R. meliloti* f1-12, lanes 2 and 7 *R. meliloti* f1-27, lanes 3 and 8 *R. meliloti* f1-40, lanes 4 and 9 *R. meliloti* f1-41, and lanes 5 and 10 *R. meliloti* MVII. Lanes 11 through 13 and lanes 14 through 16 show amplifications again done with DAF-3 and DAF-4, respectively. Lanes 11 and 14 show *R. leguminosarum* bv. *viciae* VF39. Lanes 12 and 15 *R. leguminosarum* bv. *trifolii* 1008 and lanes 13 and 16 *R. leguminosarum* bv. *phaseoli* 8002. Lane 17 displays a size marker (PhiX174 RF-DNA digested with *Hae*III; sizes in bp from top to bottom: 1353, 1078, 872, 603, 310, 281, 234, 194) All five *R. meliloti* isolates share some bands in the experiments done with DAF-4, while others are different. Two of the three *R. leguminosarum* strains (VF39 and 1008), in spite of belonging to different biovars, display nearly identical patterns in amplifications using DAF-4, whereas patterns generated with DAF-3 demonstrate that the strains are different.

cases, it is time consuming and costly to assign strains correctly to species within a genus by conventional means. Without changing the PCR-fingerprinting procedure itself, but by using a different kind of primer, the generation of species-specific (instead of strain-specific) fingerprints is possible. The method developed by Welsh and McClelland (27) makes use of conserved sequences at the ends of tRNA genes in bacteria. Two primers with a tRNA consensus sequence complementary to sequences within the genes and pointing outward, prime the amplification of DNA fragments corresponding to the regions between the clustered tRNA genes. The patterns obtained allow assignment of bacterial strains to species provided that a type strain for each species in question is available as a reference. After such a pre-classification is done, strains within groups can be compared by analysis of RAPD patterns and fragments of equal size present in different strains are more easily recognized.

The tRNA consensus fingerprinting method has been used to type species in the genera *Streptococcus* and *Enterococcus* (27) and *Pseudomonas solanacearum* (28). In the case of *Rhizobium* strains, tRNA consensus fingerprinting clearly showed the close relationship between strains of the three *R. leguminosarum* biovars, which are classified as one single species whereas the pattern of *R. meliloti* is different (Fig.3).

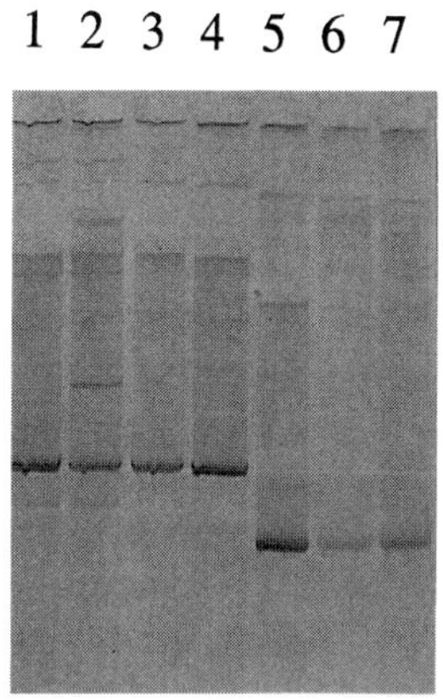

Fig. 3. PCR Fingerprinting of Rhizobium Strains with tRNA Consensus Primers. Nearly identical patterns were obtained from all *R. meliloti* strains (Lanes 1 through 4 showing *R. meliloti* strains f1-12, f1-27, f1-40 and MVII). *R. leguminosarum* members of the three biovars viciae (lane 5), trifolii (lane 6) and phaseoli (Lane 7) are indistinguishable with this technique, which is in accordance with the recent reclassification of the former three species *R. leguminosarum*, *R. trifolii*, and *R. phaseoli* as one single species containing three biovars.

8.2.6 Automated Analysis of Fingerprints

The application of the PCR for fingerprinting purposes makes the automation of the whole process possible. The generation of the fragments by PCR is already automated, the analysis of the fragment sizes now done by gel electrophoresis and staining can also be automated. Separation of the amplification products can be performed by an automated sequencer (29), allowing on-line detection of fragments once they pass the detecting laser beam. Using the A.L.F. sequencer (Pharmacia Biotech), results can be displayed as a band pattern or as traces (Fig. 4). In addition, electronic storage and management of data are directly connected with the automated detection. Once appropriate software is available, the comparison of newly obtained patterns with patterns stored in a data bank can be done by computer. Thus, screening of a large number of patterns becomes possible.

8.3 Outlook

We have attempted to give an outline of currently available DNA fingerprinting methods for the typing of bacterial strains and species. Although they will not (and should not) replace conventional techniques, they will contribute to the fast characterization and verification of bacterial strains. This could provide great advantages in many fields, especially in hygiene, food production, production of bacterial starter

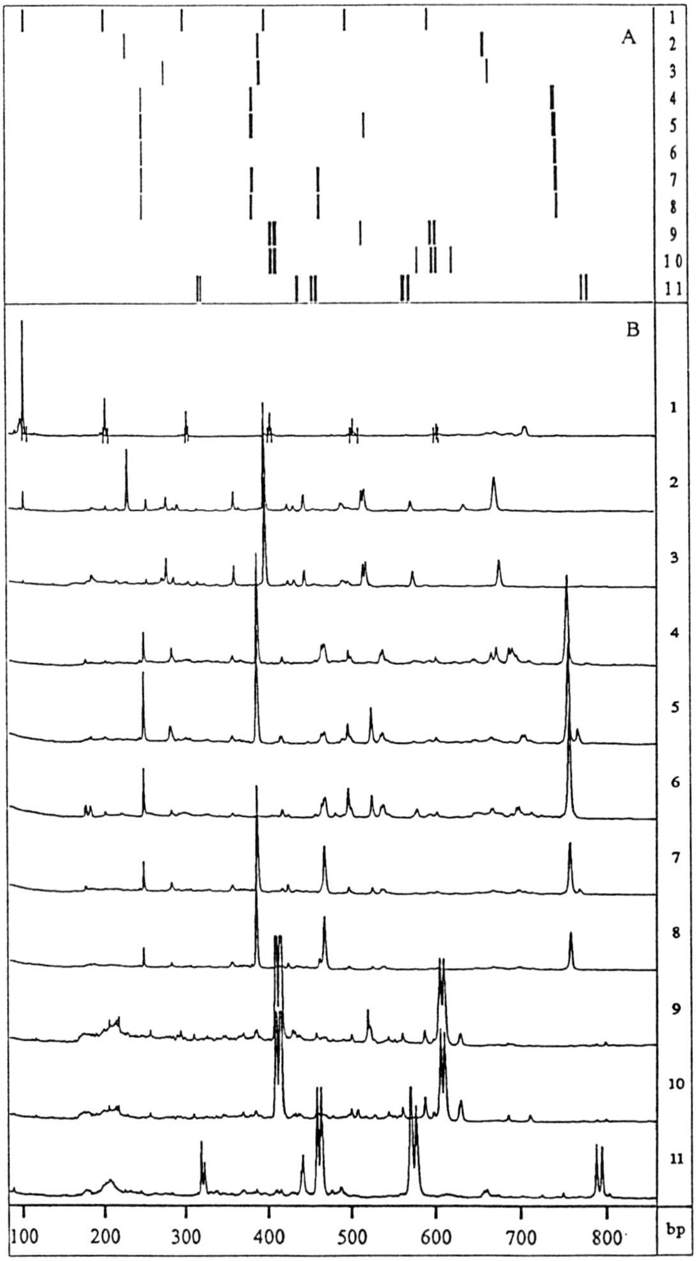

Fig. 4. Automated Analysis of Fingerprints with the A. L. F. Sequencer. Data are displayed as band pattern (A) and as traces (B). Patterns of the following *Rhizobium* strains (Primer used in parentheses) are shown: Lane 2: *R. meliloti* f1-41 (DAF-2), Lane 3: *R. meliloti* MVII (DAF-2), Lane 4: *R. meliloti* f1-12 (DAF-4), Lane 5: *R. meliloti* f1-27 (DAF-4), Lane 6: *R. meliloti* f1-40 (DAF-4), Lane 7: *R. meliloti* f1-41 (DAF-4), Lane 8: *R. meliloti* MVII (DAF-4), Lane 9: *R. leguminosarum* bv. *viciae* VF39 (DAF-4), Lane 10: *R. leguminosarum* bv. *trifolii* 1008 (DAF-4), Lane 11: *R. leguminosarum* bv. *phaseoli* 8002 (DAF-4). Lane 1 shows a 100 bp ladder used as a size marker. Strains *R. meliloti* f1-41 and MVII, giving identical patterns in amplifications with DAF-4, can be discriminated by using a different primer (DAF-2).

cultures for food processing, etc. In these cases, it is essential to have information immediately either about a contaminating organism or a commercial bacterial culture to provide a basis for early action. In the area of ecology, large numbers of isolates have to be characterized. For this purpose, PCR-fingerprinting is the method of choice, because large numbers of samples can be processed in a relatively short time, and no species-specific DNA probes are needed. Conventional techniques have great potential in the area of species identification, but some fingerprinting methods are now available which could be used for species identification in the future. Strain typing, however, could be done mainly by PCR techniques with conventional typing methods only used to verify the fingerprinting data, if necessary.

Acknowledgements

We gratefully acknowledge the technical assistance of B. Jäger and the supply of the A.L.F. data shown in Fig. 4 by P. Wiesner (Pharmacia Biotech).

8.4 References

1. Jeffreys, A.J. DNA sequence variants in γ, α, δ and β-globin genes in man. Cell *18* (1979) 1–10.
2. Wyman, A.R. and R. White. A highly polymorphic locus in human DNA. Proc. Natl. Acad. Sci. USA *77* (1980) 6754–6758.
3. Jeffreys, A.J., J.F.Y. Brookfield, and R. Semeonoff. Positive identification of an immigration test case using human DNA fingerprints. Nature *317* (1985) 818.
4. Nakamura, Y., M. Leppert, P. O'Connell, R. Wolff, T. Holm, M. Culver, C. Martin, E. Fujimoto, E. Kumlin, and R. White. Variable Number of Tandem Repeat (VNTR) Markers for Human Gene Mapping. Science *235* (1987) 1616–1622.
5. Jeffreys A.J., V. Wilson, R. Neumann and J. Kayte. Amplification of human minisatellites by the polymerase chain reaction: towards DNA fingerprinting of single cells. Nucleic Acids Res. *16* (1988) 10953–10971.
6. Grimont, F., and P.A.D. Grimont. Ribosomal ribonucleic acid gene restriction patterns as potential taxonomic tools. Ann. Inst. Pasteur/Microbiol. (Paris) *137B* (1986) 165–175.
7. Vaneechoutte, M., R. Rossau, P. De Vos, M. Gillis, D. Janssens, N. Paepe, A. De Rouck, T. Fiers, G. Claeys and K. Kersters. Rapid identification of bacteria of the Comamonadaceae with amplified ribosomal DNA-restriction analysis (ARDRA). FEMS Microbiol. Lett. *93* (1992) 227–234.
8. Welsh J. and M. McClelland. Fingerprinting genomes using PCR with arbitrary primers. Nucleic Acids Res. *18* (1990) 7213–7219.
9. Williams J.G.K., A.R. Kubelik, K.J. Livak, J.A. Rafalski and S.V. Tingey. DNA polymorphisms amplified by arbitrary primers are useful as genetic markers. Nucleic Acids Res. *18* (1990) 6531–6535.
10. Harrison S.P., L.R. Mytton, L. Skot, M. Dye and A. Cresswell. Characterization of *Rhizobium* isolates by amplification of DNA polymorphisms using random primers. Can. J. Microbiol. *38* (1992) 1009–1015.
11. Bassam B.J., G. Caetano-Anollés and P.M. Gresshoff. Fast and sensitive silver staining of DNA in polyacrylamide gels. Anal. Biochem. *80* (1991) 81–84.
12. Bassam B.J., G. Caetano-Anollés and P.M. Gresshoff. DNA amplification fingerprinting of bacteria. Appl. Microbiol. Biotechnol. *38* (1992) 70–76.
13. Sellstedt A., B. Wullings, U. Nyström and P. Gustafsson. Identification of *Casuarina-Frankia* strains by use of polymerase chain reaction (PCR) with arbitrary primers. FEMS Microbiol. Lett. *93* (1992) 1–6.
14. Welsh J., C. Pretzman, D. Postic, I. Saint Girons, G. Baranton and M. McClelland. Genomic fingerprinting by arbitrarily primed polymerase chain reaction resolves *Borrelia burgdorferi* into three distinct phyletic groups. Int. J. Syst. Bacteriol. *42* (1992) 370–377.

15. Crowhurst R.N., B.T. Hawthorne, E.H. Rikkerink and M.D. Templeton. Differentiation of *Fusarium solani* f. sp. *cucurbitae* races 1 and 2 by random amplification of polymorphic DNA. Curr. Genet. *20* (1991) 391-396.
16. Khush R.S., E. Becker and M. Wach. DNA Amplification polymorphisms of the cultivated mushroom *Agaricus bisporus*. Appl. Environ. Microbiol. *58* (1992) 2971-2977.
17. Echt C.S., L.A. Erdahl and T.J. McCoy. Genetic segregation of random amplified polymorphic DNA in diploid cultivated alfalfa. Genome *35* (1992) 84-87.
18. Welsh J., R.J. Honeycutt, M. McClelland and B. Sobral. Parentage determination in maize hybrids using the arbitrarily primed polymerase chain reaction (AP-PCR). Theor. Appl. Genet. *82* (1991) 473-476.
19. Baird E., S. Cooper-Bland, R. Waugh, M. DeMaine and W. Powell. Molecular characterization of inter- and intra-specific somatic hybrids of potato using randomly amplified polymorphic DNA (RAPD) markers. Mol. Gen. Genet. *233* (1992) 469-475.
20. Kaemmer D., R. Afza, K. Weising, G. Kahl and F.J. Novak. Oligonucleotide and amplification fingerprinting of wild species and cultivars of banana (*Musa* spec.). Bio/Technology *10* (1992) 1030-1035.
21. Klein-Lankhorst R.M., A. Vermunt, R. Weide, T. Liharska and P. Zabel. Isolation of molecular markers for tomato (*L. esculentum*) using random amplified polymorphic DNA (RAPD). Theor. Appl. Genet. *83* (1991) 108-114.
22. Caetano-Anollés G., B.J. Bassam and P.M. Gresshoff. High resolution DNA amplification fingerprinting using very short arbitrary oligonucleotide primers. Bio/Technology *9* (1991) 553-557.
23. Welsh J., C. Peterson and M. McClelland. Polymorphisms generated by arbitrarily primed PCR in the mouse: application to strain identification and genetic mapping. Nucleic Acids Res. *19* (1991) 303-306.
24. Woodward S.R., J. Sudweeks and C. Teuscher. Random sequence oligonucleotide primers detect polymorphic DNA products which segregate in inbred strains of mice. Mamm. Genome *3* (1992) 73-78.
25. Caetano-Anollés G., B.J. Bassam and P.M. Gresshoff. Primer-template interactions during DNA amplification fingerprinting with single arbitrary oligonucleotides. Mol. Gen. Genet. *235* (1992) 157-165.
26. Simon, R., B. Hötte, B. Klauke, and B. Kosier. Isolation and characterization of insertion sequence elements from gram-negative bacteria by using new broad-host-range, positive selection vectors. J. Bacteriol. *173* (1991) 1502-1508.
27. Welsh J. and M. McClelland. Genomic fingerprints produced by PCR with consensus tRNA gene primers. Nucleic Acids Res. *19* (1991) 861-866.
28. Seal S.E., L.A. Jackson and M.J. Daniels. Use of tRNA consensus primers to indicate subgroups of *Pseudomonas solanacearum* by polymerase chain reaction amplification. Appl. Environ. Microbiol. *58* (1992) 3759-3761.
29. Cancilla M.R., J.B. Powell, A.J. Hiller and B.E. Davidson. Rapid genomic fingerprinting of *Lactococcus lactis* strains by arbitrarily primed polymerase chain reaction with ^{32}P and fluorescent labels. Appl. Environ. Microbiol. *58* (1992) 1772-1775.

9 Development of Subtraction Hybridization Procedures for Generating Strain-Specific *Rhizobium* DNA Probes

J.E. Cooper and A.J. Bjourson

9.1 Introduction

The basic feature of subtraction hybridization protocols is the removal from one cell type of nucleic acid sequences which are shared with other cell types (sources of subtracter sequences), to leave only those sequences which are unique to the cell type or organism in question. Subtracter sequences may be from one or several related organisms and they can be modified in a variety of ways to permit the separation of unwanted hybrids from cell-specific sequences in the hybridization mixture. These modifications include immobilization on solid supports to facilitate separation from the mobile phase by centrifugation (1,2) and biotinylation to allow removal of unwanted hybrids by binding to avidin-coated beads (3,4) or streptavidin-phenol-chloroform extraction (5).

These techniques are limited in their capacities to generate a pool of highly enriched cell-specific sequences by their reliance on a single, partially efficient separation system and/or by an inability to amplify the small quantities of nucleic acid generated by each round of subtraction.

We describe below stages in the development of a combined subtraction hybridization and PCR amplification procedure which overcomes these problems and which has been used to generate highly discriminating DNA probes for strains within several *Rhizobium* species.

9.2 System with Biotinylated and Mercurated Subtracter DNA

Initial work was focused on improving the efficiency of separating hybrids from unique sequences in the subtraction mixture by adopting a double labelling system for the subtracter DNA. A diagram of the procedure is shown in Fig 1. In this diagram the organism from which unique DNA sequences are to be isolated is termed the probe strain, while the subtracter DNA consists of pooled total genomic DNA from

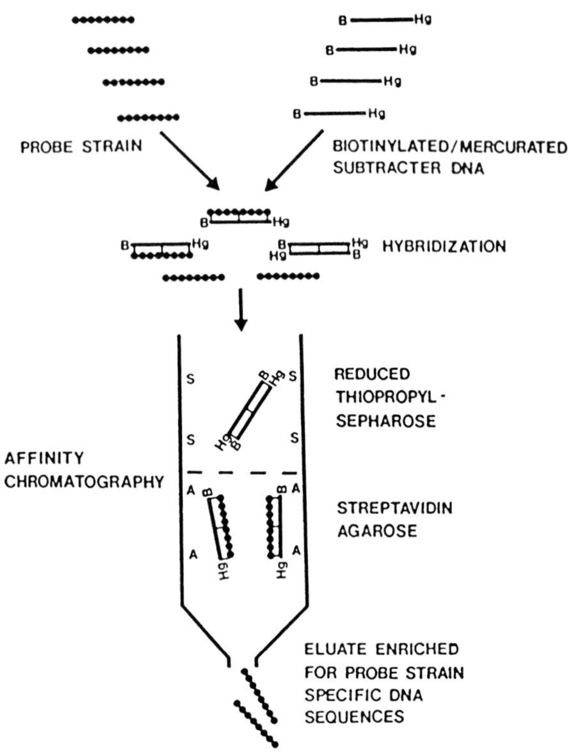

Fig. 1. Subtraction Hybridization System with Biotinylated and Mercurated Subtracter DNA.

a group of related organisms. One μg of denatured probe strain DNA is mixed with 500 μg biotinylated and mercurated subtracter DNA and hybridized at 35 °C for 48 h. Biotinylated, mercurated subtracter DNA and probe strain sequences hybridized to it are removed by two-step affinity chromatography on streptavidin agarose and thiol-Sepharose. The specificity of the sequences remaining after subtraction hybridization is assessed by dot blot hybridizations against total genomic target DNA from each strain which contributed to the pool of subtracter DNA. When this method was used to isolate a probe for one strain of *Rhizobium loti* in a group of three strains, two rounds of subtraction were sufficient to remove cross-hybridizing sequences (1).

9.3 Combined Subtraction Hybridization and PCR Amplification Procedure

Further developments were aimed at eliminating the toxic mercuration procedure and further improving the efficiency of the separation system. Additionally a means of amplifying the small quantity of strain-specific DNA remaining after subtraction was required. This resulted in the technique shown in Fig 2, in which four separation strategies (*) are used to isolate unique DNA sequences from the genome of cell type "A" after its hybridization with total genomic DNA from related cell types, "B". *Sau*3A-digested DNA from cell type "A" is ligated to a linker, denatured to single stranded form and hybridized in solution with a vast excess of subtracter DNA from cell type "B", which has been restricted, ligated to a subtracter-specific biotinylated linker, amplified by PCR to incorporate dUTP and has similarly been denatured. Subtracter DNA and "A-B" hybrids are then removed by phenol-chloroform extraction of a streptavidin-biotin-DNA complex. Nensorb chromatography of the sequences remaining in the aqueous layer captures biotinylated subtracter DNA which may have escaped removal by the phenol-chloroform treatment. Traces of contaminating subtracter DNA are removed by digestion with uracil DNA glycosylase. Finally, remaining sequences are amplified by PCR with the type "A"-specific primer, labelled and tested for specificity in dot blot hybridizations against total genomic target DNA from cell type "B". Removal of cross-hybridizing sequences is normally achieved after 1 or 2 rounds of subtraction/amplification.

9.3.1 Technical details

The information given below is generally applicable to probe generation in bacteria. The organism for which a probe is required is termed the "probe strain" and the group of related organisms, which share some homology with the probe strain, are termed "subtracter strains".

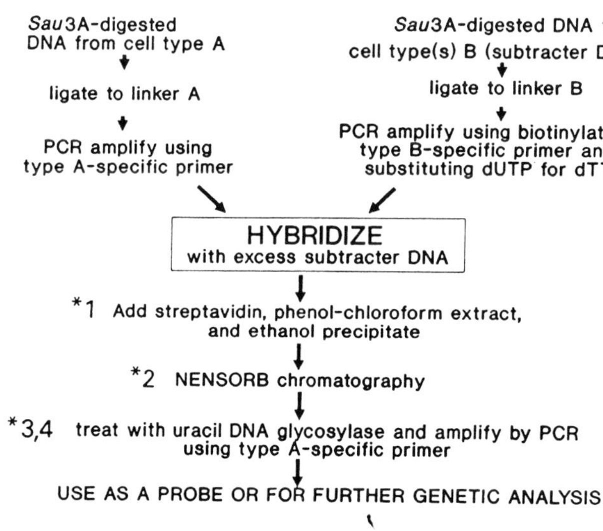

Fig. 2. Combined Subtraction Hybridization and PCR Amplification Procedure Showing the Four (*) Separation Strategies.

9.3.1.1 Isolation of DNA

Pure cultures of bacteria are grown in 50 ml conical flasks containing 25 ml of appropriate growth medium. Of each broth culture 5 ml are pelleted in a bench centrifuge and the resuspended pellets (50 μl) transferred to 1.5 ml centrifuge tubes followed by washing three times with 500 μl TE buffer (10 mM Tris.HCl, 1 mM EDTA, pH 8.0). The final pellet is resuspended in 100 μl of a solution containing 25% sucrose, 1 mg/ml lysosyme, 10 mmol/1 Tris.HCl, pH 8.0 at 37 °C for 15 min. Cells are lysed by the addition of 200 μl of lysis solution (5 mol/1 guanidine isothiocyanate, 0.1 mol/1 EDTA, pH 7.0). The lysate is mixed gently with 150 μl of 7.5 mol/1 ammonium acetate and the mixture is extracted with 500 μl of chloroform/isoamyl alcohol (24:1 v/v) by mixing and centrifuging in a microcentrifuge for 10 min. The aqueous layer is transferred to a clean 1.5 ml centrifuge tube and the DNA is precipitated by the addition of 0.54 vol 2-propanol. DNA is collected by spooling on a pipette tip, washed twice in 100 μl of 70% ethanol and then resuspended in 20 μl of TE buffer. The concentration is estimated by gel electrophoresis of 5 μl volumes. Approximately 1 μg DNA from each strain is digested with restriction endonuclease *Sau*3A, and the restriction fragments are purified with PREP-A-GENE matrix (BIORAD) as described by the manufacturer and resuspended in 20 μl of TE buffer.

9.3.1.2 Synthesis of oligonucleotides and preparation of linkers

Oligonucleotides with the following base sequences are synthesised:

TB7006 5′ HO-AGCGGATAACAATTTCACACAGGA-OH 3′

TB7007 5′ BIOTIN-CGCCAGGGUUUUCCCAGUCACGAC-OH 3′

TB7008 5′ P-GATCTCCTGTGTGAAATTGTTATCCGCT-OH 3′

TB7009 5′ P-GAUCGUCGUGACUGGGAAAACCCUGGCG-OH 3′

Oligonucleotides are resuspended in sterile TE buffer at a final concentration of 200 μM and stored in aliquots at −20 °C. Each of the appropriate oligonucleotides (5 μg) is combined to produce the linkers L-P (TB7006 and TB7008) and L-S (TB7007 and TB7009). The mixtures are heated to 65 °C and cooled slowly at room temperature to produce double stranded linkers containing 5′-phosphorylated *Sau*3A-compatible overhangs at one end. Linker L-P is ligated to *Sau*3A-digested probe strain DNA and linker L-S is ligated to similarly digested subtracter DNA. In each case *Sau*3A-digested DNA (200 ng) is mixed with 600 ng of the appropriate linker and the mixture is ligated with DNA ligase. Excess linkers are removed with PREP-A-GENE matrix by procedures described by the manufacturer and linked DNA is eluted in 20 μl of TE buffer at 50 °C.

9.3.1.3 Preparation of probe strain DNA

Probe strain DNA (1 μg) modified by ligation to linker L-P is amplified with 45 cycles of PCR. Each cycle consists of denaturation at 94 °C (1 min 20 sec), annealing at 55 °C (1 min) and DNA polymerisation at 72 °C (2 min) in an automated thermal cycler (Perkin Elmer Cetus, model 480). Reactions are performed in sterile 0.5 ml tubes with 100 μl final reaction volumes containing Tris, pH 8.3, 10 mmol/l; KCl, 50 mmol/l; $MgCl_2$, 1.5 mmol/l; gelatin, 0.01% (w/v); dNTP's, 200 μmol/l; primer TB7006, 1 μmol/l; 0.5 units of AmpliTaq DNA polymerase (Perkin Elmer Cetus). Evaporation from the tubes is prevented by addition of a 100-μl mineral oil overlay.

9.3.1.4 Preparation of subtracter DNA

Subtracter DNA from the required number of subtracter strains, modified by ligation to linker L-S, is amplified by PCR, as individual strain DNA according to the amplification conditions described for probe strain DNA. However, primer TB7007 is used instead of TB7006 and dUTP is substituted for dTTP to give a final dUTP concentration of 300 μmol/l. The efficiency of amplification is examined by submitting 10 μl volumes of each reaction to electrophoresis in composite gels (3% NuSieve agarose, 1% SeaKem agarose (w/v), FMC Bio Products). Subtracter DNA PCR products are pooled, the mineral oil is removed by extraction with an equal volume of chloroform/isoamyl alcohol (24 : 1), the aqueous phase is transferred to a Centricon 30 microconcentrator (Amicon Ltd, UK) and the mixture is concentrated by spin dialysis to yield a final volume of 25–50 μl. The volume is made up to 2 ml with 1 mmol/l EDTA and the spin dialysis step repeated until the residual volume of the mixture is approximately 25 μl. This step is repeated with another 2 ml of 1 mmol/l EDTA.

9.3.1.5 Subtraction hybridization

PCR-amplified probe strain DNA (1–5 ng) and approximately 20 μg subtracter DNA are mixed in a 0.5 ml microcentrifuge tube containing hybridization solution to give a total volume of 10 μl containing 50 mmol/l HEPES, pH 7.5; 500 mmol/l NaCl; 1 mmol/l EDTA and 0.1% SDS (final concentrations). The mixture is overlaid with 50 μl mineral oil and the DNA is denatured at 99 °C for 10 min, cooled rapidly on ice and incubated at 64 °C for 48 h to allow subtraction hybridization to take place.

9.3.1.6 Isolation of probe strain DNA sequences from the subtraction mixture

A solution (100 μl) containing 500 mmol/l NaCl, 1 mmol/l EDTA and 50 mmol/l HEPES is added and the mixture is briefly centrifuged, excess mineral oil removed and 10 μg of streptavidin is added. The tube is mixed gently at room temperature for 5 min and a further 20 μg of streptavidin is added. The mixture is extracted with an equal volume of phenol/chloroform (50:50 v/v) and centrifuged in a microcentrifuge for 10 min at 15,000 x g. The aqueous phase is transferred to a fresh tube, SDS is added to 0.1 % and the mixture is extracted twice more with phenol/chloroform, once with chloroform, and the DNA remaining in the aqueous phase is precipitated after the addition of 3 mol/l sodium acetate (0.1 vol) and 2 vol 99 % ethanol. The pellet, which may not be visible, is washed with 100 μl 70 % ethanol and redissolved in 20 μl TE buffer, pH 8.0. The deproteinized DNA is desalted with a NENSORB 20 purification cartridge (Du Pont Ltd, UK) as described by the manufacturers, and finally resuspended in 10 μl TE buffer. This subtracted DNA (5 μl), enriched for probe strain-specific sequences, is used for another subtraction cycle and 1 μl of the remaining 5 μl is used to prepare logarithmic dilutions in 9 μl aliquots of TE buffer. Of each dilution 10 μl is mixed with PCR reagents (Tris, pH 8.3, 10 mmol/l; KCl, 50 mmol/l; $MgCl_2$, 1.5 mmol/l; gelatin, 0.01 % (w/v); dNTP's 200 μmol/l) containing 15 units of uracil DNA glycosylase (Cetus) and incubated at 37 °C for 1–4 h to destroy all traces of dUTP-containing subtracter DNA. Primer TB7006 (5 μl of 200 μM, probe strain-specific) and 0.5 units of Taq polymerase are added to a final volume of 100 μl and the reaction mix is amplified for 45 cycles under the temperature conditions described above to prepare probe strain DNA. PCR products are detected by electrophoresis of 10 μl volumes of each PCR mix in composite gels.

9.4 Results

Specificity of sequences remaining after subtraction is assessed by dot blot hybridization against total genomic target DNA from strains contributing to the pool of subtracter DNA. Fig 3 shows the specificity of DNA sequences from one strain (row 1) of *Rhizobium leguminosarum* bv. *trifolii* when they were used to probe homologous target DNA (row 1) and DNA from seven other strains (rows 2–8) of the same biovar which comprised the subtracter material. Before subtraction hybridization (column A) DNA from the prospective probe strain cross-hybridized with all other strains. A small reduction in cross-hybridization with some subtracter strains was achieved after one round of subtraction/amplification (column B) and a second round generated a set of sequences which hybridized only with homologous target DNA (column C).

9.5 Conclusions

The combined subtraction hybridization and PCR amplification procedure facilitates the rapid isolation of high-specificity DNA probes from organisms for which no previous nucleic acid sequence information is available. Independent priming of

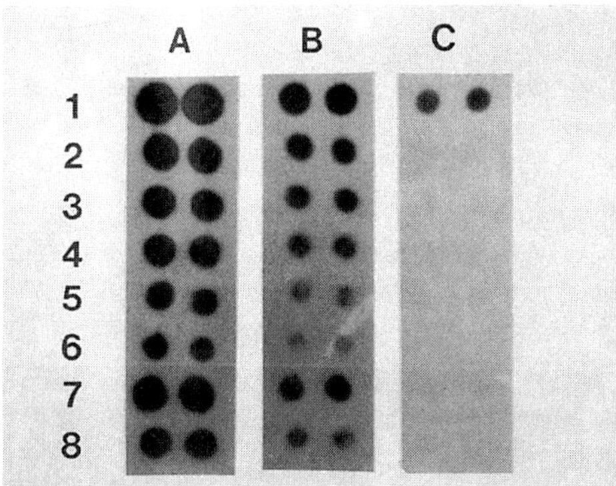

Fig. 3. Specificity of *R. leguminosarum* bv. *trifolii* Strain P3 DNA Sequences Generated by Subtraction Hybridization. ^{32}P-labelled probe strain P3 DNA submitted to zero (A), one (B), or two (C) subtraction cycles was hybridized in duplicate to 1 μg of total genomic DNA from strains P3, IDL, 1192, 1520, 1312, ANU618, LPR5035, and 14 in rows 1 to 8, respectively. For each subtraction cycle, the subtracter DNA was generated from the strains in rows 2 through 8.

subtracter DNA before the subtraction event ensures a renewable supply of this material in sufficient quantities to provide the necessary large molar excess over probe strain DNA. Specific priming of probe strain DNA permits amplification of the small quantities of strain-specific sequences which remain after each subtraction cycle.

The efficient separation of probe strain-specific sequences from probe-subtracter hybrids is dependent on a combination of four distinct strategies: phenol-chloroform extraction of biotinylated subtracter DNA, NENSORB chromatography, destruction of dUTP-containing subtracter DNA with uracil DNA glycosylase and finally, amplification of unique probe strain sequences with a specific primer in the PCR reaction. Together, these steps ensure that probe strain sequences remaining after each round of subtraction are free from even the faintest contaminating traces of subtracter DNA.

The stringency at which the subtraction is performed is an important factor to be taken into account when considering the intended use of the final sequences; low-stringency subtraction will remove some probe strain sequences which have a relatively low base sequence homology with the subtracter DNA, whereas high-stringency subtraction will remove only perfectly matched sequences.

9.6 References

1. Bjourson, A.J. & Cooper, J.E. Isolation of *Rhizobium loti* strain-specific DNA sequences by subtraction hybridization. Appl. Environ. Microbiol. *53* (1988) 1705–1707.
2. Scott, M.R.D., Westphal, K.W. & Rigby, P.W.C. Activation of mouse genes in transformed cells. Cell *34* (1983) 557–567.

3. Straus, D., & Ausubel, F.M. Genomic subtraction for cloning DNA corresponding to deletion mutants. Proceedings of the National Academy of Sciences *87* (1990) 1889–1892.
4. Sun, T.P., Goodman, H.M. & Ausubel, F.M. Cloning of the *Arabidopsis* GA1 locus by genomic subtraction. The Plant Cell *4* (1992) 119–128.
5. Sive, H.L. & St. John, T. A simple subtractive hybridization technique employing photoactive biotin and phenol extraction. Nucleic Acids Research *16* (1988) 10937.
6. Bjourson, A.J., Stone, C.E. & Cooper, J.E. Combined subtraction hybridization and polymerase chain reaction amplification procedure for the isolation of strain specific *Rhizobium* DNA sequences, Appl. Environ. Microbiol. *58* (1992) 2296–2301.

10 Molecular Characterization and Detection of the Actinomycete *Frankia* in the Environment

Pascal Simonet, Sylvie Nazaret and Philippe Normand

10.1 Introduction

Microbial ecology has long been hampered by the numerous problems encountered in isolating and culturing most of the microorganisms present in the environment (it is estimated that less than 20% of the extant microorganisms have been discovered) (1). Moreover, for the few of them which are routinely maintained in pure culture, problems exist in classifying, identifying and characterizing the different strains and a need exists for efficient methods to follow microbial strains directly in the environment. This would permit us to eliminate the time-consuming isolation step which creates bias by selecting particular phenotypes.

In this paper, we discuss how advances in molecular biology have offered a range of efficient new tools for each of the domains of microbial ecology. This will include DNA-DNA hybridizations which are now recognized as the standard methods in bacterial taxonomy permitting the characterization of genomic species. These results are now confirmed by phylogenetic analyses based on sequence data of ribosomal genes which also allow us to specifically identify microorganisms. Finally, using DNA hybridization or PCR techniques, a rapid detection and identification of microorganisms in complex habitats such as water, sediment or soil, without prior isolation and culturing is possible.

This article will focus on the nitrogen-fixing procaryote, *Frankia*, symbiont of a large number of woody dicotyledonous plants. This actinomycete belongs to the category of microorganisms that are difficult to isolate and for which culture is very tedious (24 to 48 h generation time). This limiting factor has acted as a barrier in understanding the mechanisms of nitrogen fixation and symbiosis by *Frankia* but also in evaluating natural *Frankia* population diversity in addition to the ecology of the microorganism as a saprophyte or a symbiont.

Ecological data concerning *Frankia* strains is useful for applied purposes. The ability of this actinomycete to establish a symbiosis with the so-called actinorhizal plants permits them to grow on nitrogen-poor soils or in disturbed environments which has led to their use for reforestation, reclamation of mine spoils, dune stabilization and timber production (2). Large scale plantations of *Casuarina equisetifolia* have been successfuly reported in Senegal and China where millions of hectares had been planted to stabilize coastlines. Experimental data indicate that the growth of these plants in adverse environments can be optimized by nodulation with effective strains. This requires isolation of a large range of compatible strains which have to be tested for their efficiency in fixing nitrogen, their ability to develop a symbiosis and to adapt to the new environmental conditions. These applied purposes are compatible with more fundamental ones on the ecology of *Frankia* in soil and in nodules and also require methods to classify, characterize, identify and detect strains of this actinomycete.

10.2 Taxonomy

10.2.1 DNA-DNA hybridization data

Since the first successful isolation of a *Frankia* strain in 1978 (3) different criteria have been used to characterize the genus *Frankia* and classify isolates obtained from several plant species from different geographical areas. These include; the typical *Frankia* morphology composed of hyphae, vesicles and multilocular sporangia; biochemical parameters such as a cell wall of type III and phospholipid of type I; physiology and infectivity on host-plants. Several of these criteria used alone or in combination were found to be efficient in separating the genus *Frankia* from most of the actinomycetes but, generally failed to differentiate clearly between isolates.

This was also the case with inoculation tests which defined four overlapping infectivity groups based on the ability of strains to develop a symbiosis with actinorhizal plants (4). This grouping of strains was the first attempt to identify isolates as *Frankia* strains and to classify them. However, the technique did not allow the classification of the numerous unisolated strains and was inadequate for most ecological purposes as it was inefficient in differentiating strains belonging to the same group.

In the 1970s, a method based on DNA/DNA hybridization of the whole genome was developed for taxonomic purposes. This method became the standard method recommended by most bacteriologists to define species (1). Comparing results obtained for various taxonomic groups by several different taxonomic criteria suggested the 70% reassociation level at optimal temperature with a ΔTm below 5 °C as the threshold above which two strains should be considered as belonging to the same genomic species. This technique was applied on more than 40 *Frankia* strains (5) and resulted in the delineation of at least nine genomic species, three of them belonging to the *Alnus* infectivity group, five to the *Elaeagnus* infectivity group and one to the *Casuarina* infectivity group. Nine strains did not fall into any of these genomic groups and thus may represent one or more additional groups. Using another set of strains, Akimov et al. (6), reported the existence of five genomic species among strains infective on *Alnus* and four among strains infective on *Elaeagnus* confirming that a high degree of heterogeneity exists amongst *Frankia* strains.

10.2.2 Sequencing of 16S rDNA genes

DNA-DNA hybridization methods are time consuming and usually applied only to fast-growing microorganisms. This has led to the development of new methods to confirm DNA-DNA hybridizations. 16S ribosomal RNA sequence analysis has been particularly useful in inferring evolutionary relationships between eubacteria, archaebacteria, and eucaryotic organisms (7). The small subunit rRNA (also known as 16S gene) which is universally present in all living forms was found to exhibit conserved regions that can be used to study distantly related species. Regions of higher variability are useful for the analysis of closely related species (7) (Figure 1).

Techniques related to 16S rRNA sequencing, including oligonucleotide cataloguing or direct sequencing has demonstrated that *Frankia* strains show a high degree of relatedness with the genera *Geodermatophilus*, and *Blastococcus*. These three genera are classified as members of the *Frankiaceae* family (8). The hypervariable regions of the 16S ribosomal gene were sequenced to estimate the phylogenetic relationships between the different genomic species grouped in the *Frankia* genus. This was done for 35 strains (9) chosen from 8 of the 9 genomic species of *Frankia* (5), by using the polymerase chain reaction followed by a direct sequencing approach. This strategy led to the rapid determination of 16S rDNA sequences of the strains available in pure culture and was also successfully applied to the sequencing of non-culturable strains. An example was an *Alnus* infective strain called Sp+ and for a group of strains infecting plants belonging to the *Coriariaceae* family (10). The strategy consisted of extracting DNA from the nodule and using it as template for the polymerase chain reaction. The use of specific primers permitted the amplification of the *Frankia* rDNA sequence from a background of non-target DNA. Evolutionary distance values were drawn from sequence variations and permitted the construction of a phylogenetic tree (Figure 2).

The results indicate that two major subdivisions can be distinguished separating the genomic species infective on *Elaeagnus* from those isolates infective on *Alnus* or *Casuarina*. Moreover, the *Elaeagnus* group, considered to be a very heterogenous group when using phenotypic criteria such as fatty acid or isozyme patterns was found to exhibit less diversity than the *Alnus* infective strains when using 16S rDNA homologies. The determined partial 16S rDNA sequence of strain Sp+ permitted the positioning of this new strain in the phylogenetic tree, close to the *F. alni* species. *Coriaria* endophytes appear to belong to a unique *Frankia* lineage with *Casuarina*-infective strains as their closest neighbours. The difference found in the partial sequences amongst the *Coriaria* endophytes would indicate that these strains belong to two or more closely related genomic species. For strains isolated from *Casuarina* root nodules, ribosomal sequence data are in complete agreement with inoculation tests. These strains had been previously classified into two subgroups with respect to their ability to reinfect their original hosts (11). The phylogenetic tree confirmed this classification, the typical strains (infective on their original host *Casuarina*) branching with the *Alnus* group, whereas atypical ones (non infective on their original host *Casuarina* but infective on *Elaeagnus*) branched with *Elaeagnus* infective strains.

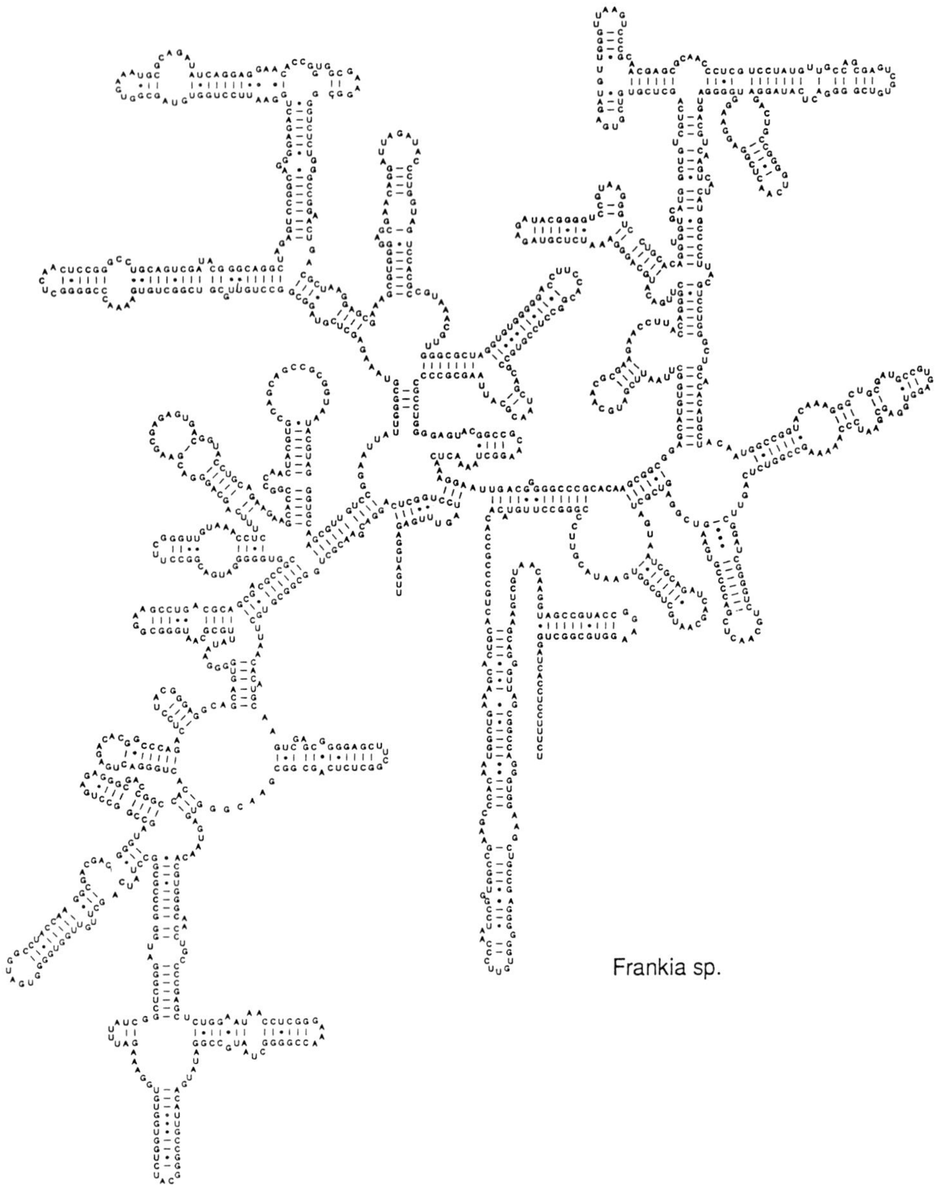

Fig. 1. Secondary structure of the 16S ribosomal rRNA gene of *Frankia* sp. strain ORS020606, kindly done by R.R. Gutell (Dept. of molecular, cellular and developmental biology, University of Colorado, Boulder, Colorado, USA).

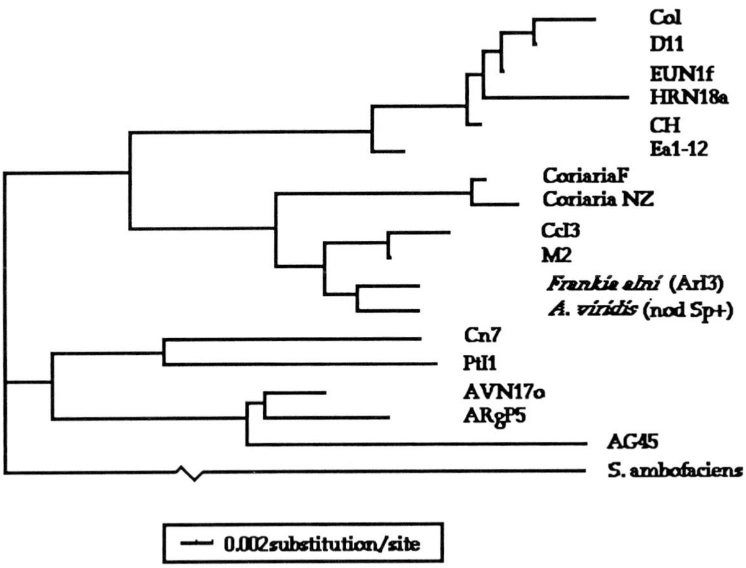

Fig. 2. Neighbour-joining tree made on the partial sequences of the 16S ribosomal gene of *Frankia* strains and un-isolated endophytes from nodules.

10.3 Characterization of *Frankia*

10.3.1 Conventional Techniques

Before the PCR revolution, a set of biochemical or DNA-based methodologies were used to identify, characterize and estimate the diversity of *Frankia* populations. These techniques included total protein (12, 13, 14), isoenzyme patterns (15, 16, 17), fatty acid (18) or sugar analysis (19). We demonstrated previously how DNA-DNA hybridization methods (5, 6, 20, 21) were used in taxonomy while RFLP pattern analysis (21, 22, 23, 24, 25, 26) were used for characterization purposes. The ability to differentiate strains harbouring a small cryptic plasmid allowed an ecological study in which plasmid-bearing *Frankia* strains were detected *in situ* by hybridization of nodule extracted DNA with the plasmid probe (27, 28).

10.3.2 Sequence Based Characterization

The use of PCR has permitted the sequencing of part of the 16S ribosomal gene for numerous *Frankia* strains whether isolated or not. Sequence comparison can be used to estimate evolutionary relationships between microorganisms and will play a role in the classification of the numerous *Frankia* strains, not available in pure culture. Moreover, the sequence data can also be used to define DNA fragments that can be used to identify the genus *Frankia*, a genomic species or a particular strain(s).

For example, sequence comparisons for ribosomal genes and also for more specific genes such as *nif* (nitrogen fixation) permitted Simonet et al. (29), and Simon et al. (30) to define on a molecular basis, the strains belonging to genus *Frankia*. This allowed the separation of *Frankia* from closely related actinomycetes of the *Frankiaceae* family such as *Geodermatophilus*. Fifteen different sequences were delineated among the 35 *Frankia* strains studied, resulting in a grouping of strains which correlated with that based on total DNA/DNA homologies. Strains belonging to the same genomic species have the same partial 16S rDNA sequences, while strains belonging to different genomic species exhibited some sequence differences (with the exception of the genomic species 4 and 5 which were undistinguishable). In conclusion, most of the genomic species can be characterized by a specific 16S rDNA sequence, providing a very useful tool to rapidly characterize new isolates or unisolated endophytes (eg. *Coriaria*-infective strains). The equivalence between identity of the 16S ribosomal rDNA sequence and belonging to the same genomic species (i. e. sharing more than 70% DNA/DNA homology) will probably become a generally accepted rule by microbial taxonomists (31).

10.3.2.1 Intergenic Spacers

Characterization at the strain level is difficult to achieve based on 16S ribosomal sequences and led us to investigate more variable regions. This was done by amplifying and sequencing the intergenic spacers (IGS) between 16S and 23S *rrn* genes and *nifH* and *nifD* which are postulated to be subject to less selection pressure. (Figure 3). Important variations in base composition and length were found between strains of different *Frankia* genomic species (unpublished results) confirming the usefulness of these DNA regions. However, the technique failed to differentiate closely related *Casuarina*-infective strains (32). This indicates that the evolution of the intergenic regions is only slightly faster compared to their conserved flanking sequences and underlines the need to develop other methods such as the hybridization-subtraction technique (see chapter 9, this volume) when closely related strains have to be differentiated.

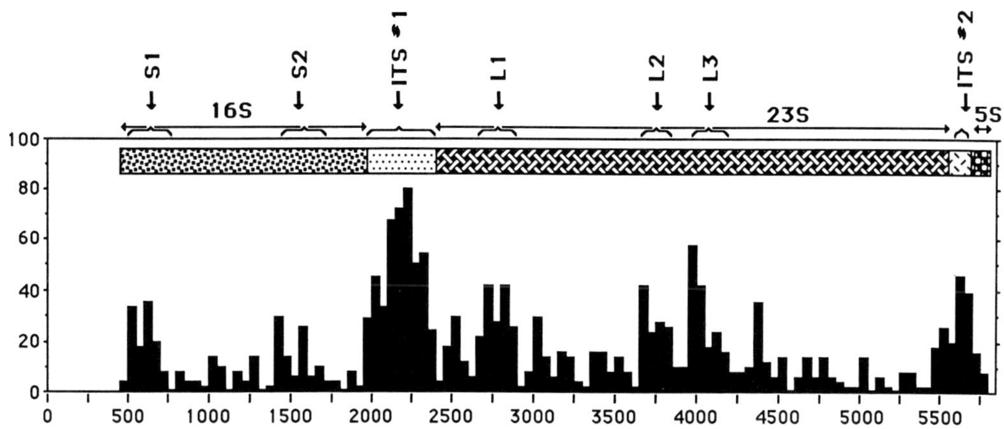

Fig. 3. Percent divergence of 50 nucleotide (nt) stretches in the ribosomal genes of *Frankia* and *Streptomyces*.

10.3.2.2 PCR/RFLP

Sequence determination appears to be the way of choice to compare and classify microorganisms. However, even with the use of PCR and direct sequencing protocols, this strategy remains time consuming and expensive when numerous strains have to be characterized. Rapid methods based on sequence information to characterize strains without determining the complete nucleotide sequence have been developed. A protocol combining PCR amplification of a target region and restriction analysis of the PCR products has proven to be very useful in obtaining information on microbial diversity. Maggia et al (33) were able to analyze the genetic diversity of more than 60 isolates, originating from *C. equisetifolia* in West Africa by comparing the restriction patterns corresponding to the intergenic spacer between the 16S and 23S *rrn* genes. In fact, the technique can be considered as a preliminary step for screening a large number of strains prior to more detailed analysis.

The RAPD technique based on the use of short primers to amplify randomly determined fragments possesses numerous advantages for characterization purposes (See Chapter 8, this volume). This was demonstrated by Sellstedt et al. (34) in characterizing patterns obtained after RAPD amplification of *Frankia* strains infective on *Casuarina* which were undistinguishable by different techniques. One possible drawback of the RAPD/PCR technique is interference from plant genomes, making it in theory a less promising technique than targeted PCR of the intergenic spacer.

10.4 Detection and Enumeration

The fact that *Frankia*, a free living actinomycete from soil can establish a specific symbiosis with some plants makes it a good model for ecological studies. Because of the physiological properties of *Frankia,* resulting in slow-growing isolates and numerous unisolated strains, there is a need for detection methods which could be applied in situ. Molecular techniques can be complementary and useful in detecting and enumerating the *Frankia* cells in nodules and also in the soil.

10.4.1 Detection of *Frankia* in Actinorhizae

A first set of experiments was conducted to study the competition between two *Frankia* strains for the nodulation of the roots of two host plant species (35). The seedlings were inoculated when the shoots had two leaves with inocula consisting of either single strains as control or an equal mixture of the two strains. The nodules were harvested 2 months after inoculation, rinsed with sterile distilled water and each nodule lobe was individually crushed before being treated according to the enzymatic protocol developed for extracting DNA from *Frankia* strains. Cells debris were sedimented by centrifugation and DNA present in the supernatant used in the PCR procedure without further purification. Two 20 mer-oligonucleotides, complementary to highly conserved sequences inside the *nif*H gene were used as PCR primers in order to amplify microsymbiont DNA extracted from actinorhizae. PCR products were subsequently hybridized using two 15 mer-oligonucleotides as probes specific for each of the strains. For numerous nodules, a positive signal was obtained for only one probe indicating that only one *Frankia* strain was responsible for the formation of the nodule. This suggested that the two strains did not possess the same ability in establishing a relationship with both host plants, one strain being more competitive than the other. This indicates that the nodulation capacity is related to the strain properties and not dependent on the host-plant. Some nodules exhibited signals of similar intensity suggesting the presence of more than one *Frankia* strain in the same nodule. However, an alternative hypothesis, is the possible presence of some *Frankia* cells that originated from the inoculum remaining trapped on the surface of the nodule. This last hypothesis was recently confirmed by *in situ* hybridizing sections of nodules with the strain-specific probes, which localised the presence of only one strain in a given nodule (Prin et al., (36) in press).

Hybridization experiments on PCR amplified products were also used to rapidly determine the identity of *Casuarina* strains originating from West Africa. These were identified in nodules by hybridizing PCR products with a specific probe complementary to sequences between *nifH* and *nifD* genes (33). Ribosomal RNA molecules can also be hybridized with specific oligonucleotide probes, such a strategy was developed by a Dutch group to detect *Frankia* strains in nodules and in the soil (37).The *nifDK* IGS is also useful for population analysis in nodules. This IGS ranges from about 200 to about 400 nucleotides and is variable. It has a phylogeny congruent to that of 16S rDNA (38). This region has been used for studying the distribution of *Frankia* strains infective on two *Elaeagnus* species in two soils.

10.4.2 Direct Detection of *Frankia* Present in the Soil

Two strategies have been developed to recover DNA from complex and heterogenous environments such as soil. The first approach consists in separating microorganisms from other soil components, the bacterial suspension is subsequently treated to extract DNA or RNA according to established protocols. The second strategy involves *in situ* extraction of nucleic acids from microorganisms, the lysis being conducted directly on the environmental samples.

Evaluating both techniques Hahn et al. (39) found that total yield of RNA was generally much lower when bacteria were first separated from soil. Purification to remove humic acids was a problem. Purification also remains the main problem when the direct extraction method is used. DNA solutions are generally contaminated with humic acids which can prevent the success of subsequent experimental work, especially enzymatic treatments such as PCR amplification or restriction with endonucleases which can also hamper hybridization by reducing its specificity (27). Various purification procedures have been described, e. g. successive phenol-chloroform extractions (39, 40), potassium acetate precipitation followed by electrophoresis and electroelution of the DNA (41) or use of commercial purification kits such as Elutip-D (Scleicher and Schuell, Dassel, FRG) columns (42) or Gene Clean products (Braunschweig, FRG).

A protocol developed in our laboratory in which each step was optimized, including lysis of cells, DNA recovery, purification and PCR amplification (42) proved to be successful. Cells were lysed by physical disruption based on repeated sonication, microwave heating and thermal shocks. The recovery of the DNA was achieved by successive washes and centrifugation while the purification was achieved by passage through several Elutip d columns. The PCR amplification protocol was also improved in order to amplify one target molecule by using "booster conditions"; lower denaturation temperatures and addition of formamide. By using this protocol, specific detection was routinely obtained for inoculated bacteria when inocula ranged from 10^7 to 10^3 bacteria. An application of the MPN calculation method to PCR results, permitted the assessment of the validity of the method. A good correlation was found between the number of cells inoculated and that detected after treatment of the soil sample. This allowed us to estimate the indigenous population of *Frankia* at 0.2×10^5 cells per g of soil, which is in agreement with results obtained by other authors (39). By restricting the specificity of the primers to particular *Frankia* genotypes the technique will be applied to study their natural distribution in soil and in nodules collected in alder stands (43).

Using a similar approach based on PCR and MPN, Hilger and Myrold (41) investigated the effect of host and non-host rhizosphere and liming on *Frankia* populations. *Frankia* specific oligoprobes were used restricting the study to the genus level and permitted the authors to conclude that these two factors did not affect the size of the *Frankia* population.

10.5 Conclusion

The development of nucleic acids techniques has provided a set of new tools for microbial ecology. The appearance of new methods such as PCR has increased the range of microorganisms, including some recalcitrant strains that can be characterized. Molecular techniques are now available to characterize, identify, detect and enumerate strains or groups of strains directly in their environment.

Several total or partial rRNA/rDNA sequences are now available for the genus *Frankia*. This permits a better understanding of the existing relationships between *Frankia* and other actinomycete genera (8) and among the different genomic species defined by the DNA-DNA hybridization method (5, 9). These data make *Frankia* one of the best known microbial taxa at the evolutionary level. However, the total intrageneric diversity may be underestimated due to the difficulties encountered in the isolation of *Frankia* from several actinorhizal plants or directly from soil. However, the extensive use of PCR techniques with template DNA directly extracted from the environment will contribute, to estimate the extent of this diversity and to correlate it to environmental parameters. Some technical improvements are still required to distinguish very closely related strains and to purify DNA extracted from highly contaminated sources. When available these techniques will provide new ways to study the behaviour of *Frankia* strains in their natural habitats, including soil and nodules.

10.6 References

1. Wayne, L.G., D.J. Brenner, R.R. Colwell, P.A.D. Grimont, O. Kandler, M.I. Krichersky, L. H. Moore, W.E.C. Moore, R.G.E. Murray, E. Stackebrandt, M.P. Starr, and H.G. Trüper. Report of the ad hoc committee on reconciliation of approaches to bacterial systematics. Int. J. Syst. Bacteriol. *37* (1987) 463–464.
2. Diem, H.G., and Y.R. Dommergues. Current and potential uses and management of *Casuarinaceae* in the tropics and subtropics. In: The biology of *Frankia* and actinorhizal plants. Schwintzer C.R. and J.D. Tjepkema (Eds.). Academic press. 1990, pp. 317–342.
3. Callaham, D., P. Del Tredici, and J.G. Torrey. Isolation and cultivation in vitro of the actinomycete causing root nodulation in *Comptonia*. Science *199* (1978) 899–902.
4. Baker, D.D. Relationships among pure cultured strains of *Frankia* based on host-specificity. Physiol. Plant. *70* (1987) 245–248.
5. Fernandez, M.P., H. Meugnier, P.A.D. Grimont, and R. Bardin. Deoxyribonucleic acid relatedness among members of the genus *Frankia*. Int. J. Syst. Bacteriol. *39* (1989) 424–429.
6. Akimov, V.N., S.V. Dobritsa, and O.S. Stupar. Grouping of *Frankia* strains by DNA-DNA homology: how many genospecies are in the genus *Frankia*? In: Proceedings of the fifth international symposium

on nitrogen fixation with non-legumes. Polsinelli, M., R. Materassi, and M. Vincenzini (Eds.). Kluwer Academic Publishers. Dordrecht/Boston/London. 1991, pp. 635-636.
7. Woese, C.R. Bacterial evolution. Microbiol. Rev. *51* (1987) 221-271.
8. Hahn, D., M.P. Lechevalier, A. Fischer, and E. Stackebrandt. Evidence for a close phylogenetic relationship between members of the genera *Frankia*, *Geodermatophilus*, and "*Blastococcus*?" and emendation of the family *Frankiaceae*. System. Appl. Microbiol. *11* (1989) 236-242.
9. Nazaret, S., B. Cournoyer, P. Normand and P. Simonet. Phylogenetic relationships among *Frankia* genomic species determined by use of amplified 16S rDNA sequences. J. Bacteriol. *173* (1991) 4072-4078.
10. Nick G., E. Paget, P. Simonet, A. Moiroud, and P. Normand. The nodular endophytes of *Coriaria* spp. form a distinct lineage within genus *Frankia*. Molecular Ecology *1* (1993) 175-181.
11. Nazaret, S., P. Simonet, P. Normand, and R. Bardin. Genetic diversity among *Frankia* isolated from *Casuarina* nodules. Plant Soil *118* (1989) 241-247.
12. Benson, D.R., and D. Hanna. *Frankia* diversity in an alder stand as estimated by sodium dodecyl sulfate-polyacrylamide gel electrophoresis of whole-cell proteins. Can. J. Bot. *61* (1983) 2919-2923.
13. Benson, D.R., S.E. Buchholtz, and D. Hanna. Identification of *Frankia* strains by two dimensional polyacrylamide-gel electrophoresis. Appl. Environ. Microbiol. *47* (1984) 489-494.
14. Gardes, M., and M. Lalonde. Identification and subgrouping of *Frankia* strains using sodium dodecyl sulfate polyacrylamide gel electrophoresis. Physiol. Plant. *70* (1987) 237-244.
15. Puppo, A., L. Dimitijevic, H.G. Diem, and Y.R. Dommergues. Homogeneity of superoxide dismutase patterns in *Frankia* strains from *Casuarinaceae*. FEMS Microbiol. Lett. *30* (1985) 43-46.
16. Gardes, M., J. Bousquet, and M. Lalonde. Isozyme variation among 40 *Frankia* strains. Appl. Environ. Microbiol. *53* (1987) 1596-1603.
17. Benoist, P., and J. Schwencke. Native agarose-polyacrylamide gel electrophoresis allowing the detection of aminopeptidase, dehydrogenase, and esterase activities at the nanogram level: enzymatic patterns in some *Frankia* strains. Anal. Biochem. *187* (1990) 337-344.
18. Simon, L., S. Jabaji-Hare, J. Bousquet, and M. Lalonde. Confirmation of *Frankia* species using cellular fatty acid analysis. System. Appl. Microbiol. *11* (1989) 229-235.
19. St-Laurent, L., J. Bousquet, L. Simon, and M. Lalonde. Separation of various *Frankia* strains in the *Alnus* and *Elaeagnus* host specificity groups using sugar analysis. Can. J. Microbiol. *33* (1987) 764-772.
20. An, C.S., W.S. Riggsby, and B.C. Mullin. Relationships of *Frankia* isolates based on deoxyribonucleic acid homology studies. Int. J. Syst. Bacteriol. *35* (1985a) 140-146.
21. Bloom, R.A., B.C. Mullin, and R.L. Tate. DNA restriction patterns and DNA-DNA solution hybridization studies of *Frankia* isolates from *Myrica pensylvanica* (Bayberry). Appl. Environ. Microbiol. *55* (1989) 2155-2160.
22. An, C.S., W.S. Riggsby, and B.C. Mullin. Restriction patterns analysis of genomic DNA of *Frankia* isolates. Plant Soil. *87* (1985b) 43-48.
23. Simonet, P., P. Normand, A. Moiroud, and M. Lalonde. restriction enzyme digestion patterns of *Frankia* plasmids. Plant Soil. *87* (1985) 49-60.
24. Dobritsa, S.V. Restriction analysis of the *Frankia* spp. genome. FEMS Microbiol. Lett. *29* (1985) 123-128.
25. Normand, P., P. Simonet, and R. Bardin. Conservation of *nif* sequences in *Frankia*. Mol. Gen. Genet. *213* (1988) 238-246.
26. Nittayajarn, A., B.C. Mullin, and D.D. Baker. Screening of symbiotic *Frankiae* for host-specificity by restriction fragment length polymorphism analysis. Appl. Environ. Microbiol. *56* (1990) 1172-1174.
27. Simonet, P., N. Thi Le, E. Teissier du Cros, and R. Bardin. Identification of *Frankia* strains by direct DNA hybridization of crushed nodules. Appl. Environ. Microbiol. *54* (1988) 2500-2503.
28. Simonet, P., N. Thi Le, A. Moiroud, and R. Bardin. Diversity of *Frankia* strains isolated from a single alder stand. Plant Soil *118* (1989) 13-22.
29. Simonet, P., M-C Grosjean, A.K. Misra, S. Nazaret, B. Cournoyer, and P. Normand. *Frankia* genus-specific characterization by polymerase chain reaction. Appl. Environ. Microbiol. *57* (1991) 3278-3286.
30. Simon, L., and M. Lalonde. Design of a *Frankia* specific probe. In: Eight international conference on *Frankia* and actinorhizal plants. September 2-3-4, Lyon, 1991.

31. Fox G.E., J.D. Wisotzkey and P. Jurtschuk. How close is close: 16S rRNA sequence identity may not be sufficient to guarantee species identity. Int. J. Syst. Bact. *42* (1992) 166–170.
32. Rouvier, C., S. Nazaret, M.P. Fernandez, B. Picard, P. Simonet, and P. Normand. *rrn* and *nif* intergenic spacers and isoenzyme patterns as tools to characterize *Casuarina*-infective *Frankia* strains. Acta Oecologica *13* (1992) 487–495.
33. Maggia, L, S. Nazaret, and P. Simonet. Molecular characterization of *Frankia* isolates from *C. equisetifolia* root nodules harvested in West Africa (Senegal and Gambia). Acta Oecologica *13* (1992) 453–461.
34. Sellstedt, A., B. Wullings, U. Nyström, and P. Gustafsson. Identification of *Casuarina-Frankia* strains by use of polymerase chain reaction (PCR) with arbitrary primers. FEMS Microbiol. Lett. *93* (1992) 1–6.
35. Simonet, P., P. Normand, A. Moiroud, and R. Bardin. Identification of *Frankia* strains in nodule by polymerase chain reaction products with strain specific oligonucleotide probes. Arch. Microbiol. *153* (1990) 235–240.
36. Prin, Y., F. Mallein-Gerin, and P. Simonet. Identification and localization of *Frankia* strains in *Alnus* nodules by in situ hybridization of *nif*H mRNA with strain specific oligonucleotide probes. J. Exp. Bot. (1993) (in press).
37. Hahn, D., E. Starrenburg, and A.D.L. Akkermans. Oligonucleotide probes that hybridize with rRNA as a tool to study *Frankia* strains in root nodules. Appl. Environ. Microbiol. *56* (1990) 1342–1346.
38. Jamann, S., Fernandez M.P. and Normand P. Typing of *Frankia* strains using PCR/RFLP of the *nifD-nifK* IGS. Molecular Ecology, (1993) (in press).
39. Hahn, D., R. Kester, E. Starrenburg, and A.D.L. Akkermans. Extraction of ribosomal RNA from soil for detection of *Frankia* with oligonucleotide probes. Arch. Microbiol., *154* (1990) 329–335.
40. Nazaret, S. Caractérisation des souches de *Frankia* symbiotes de *Casuarina*; mise au point de marqueurs génétiques. Ph.D. thesis. Université Claude Bernard LYON I, Villeurbanne, 1991.
41. Hilger, A.B., and D.D. Myrold. Method for extraction of *Frankia* DNA from soil. In: Modern techniques in soil ecology. D. A. Crossley et al., (Eds.). (1990) (in press).
42. Picard, C., C. Ponsonnet, E. Paget, X. Nesme, and P. Simonet. Detection and enumeration of bacteria in soil by direct DNA extraction and Polymerase Chain Reaction. Appl. Environ. Microbiol. *58* (1992) 2717–2722.
43. Simonet, P., M. Bosco, C. Chapelon, A. Moiroud, and P. Normand. Characterization of Sp+ and Sp- *Frankia* populations and distribution in soil and actinorhizae. (In preparation).

11 Molecular Ecology of Filamentous Actinomycetes in Soil

Peter Marsh and Elizabeth M. H. Wellington

11.1 Introduction

Actinomycetes play an important role in soil decomposition processes as they produce a wide range of hydrolytic enzymes and some species have been implicated in the breakdown of lignin and other recalcitrant polymers (1, 2). The saprophytic groups have been recovered from a range of terrestrial and aquatic habitats and many are capable of colonizing the rhizosphere (3, 4). Ability to colonize the rhizosphere may depend on the possession of antagonistic properties aimed at inhibition of other competing bacteria such as bacilli and pseudomonads. Streptomycetes may have an added advantage in being able to spread in dry soils and there is evidence that they can grow more prolifically in soil with a matric potential around -300 kPa. Field experiments done under wetter conditions have indicated that *Pseudomonas fluorescens* showed more extensive colonisation of wheat root rhizosphere than did a streptomycete (5). In this study the streptomycete was selected on the basis of its ability to inhibit root pathogens (6).

Actinomycetes, and streptomycetes in particular, are one of the most important bacterial groups for antibiotic production and have been exploited by the pharmaceutical industry during the past forty years for large scale production of clinical and veterinary drugs. However, it is members of the genus *Pseudomonas* which have been most frequently implicated in the production of antibiotics for antagonisms in the rhizosphere (7, 8, 9). Members of the genus *Streptomyces* have been used as inoculants to inhibit fungal root pathogens (10) and many actinomycetes produce siderophores which may play a role in inactivating deleterious rhizobacteria.

There is a considerable amount of circumstantial evidence that streptomycetes and other actinomycetes colonizing the rhizosphere are beneficial to plants. Practices such as liming and chitin amendments of soil often result in marked increases in the actinomycete population which has been correlated with the reduction in incidents of plant disease associated with soil borne plant pathogens (11). Microbial inoculants may also be affected by antagonism through antibiosis resulting from actinomycete activity in soil. Resistance to streptomycin in Gram-negative bacteria such as *Rhizobium* and *Bradyrhizobium* species has been linked to antagonistic activity of actinomycetes in tropical soils (12, 13).

Various studies of actinomycete populations in soil have focused on selective isolation and enumeration by culturing (14). Members of the genus *Streptomyces* appear to be the most numerous actinomycetes in a wide range of soils sampled. However, slow growth of many groups often results in poor competition on isolation plates with other soil bacteria such as bacilli and pseudomonads. In addition conditions conducive to germination of actinomycete spores may not prevail in isolation media and cause problems in detection. Accurate quantification of mycelial bacteria is difficult when using culturing techniques as a single colony may have arisen from a single spore, hyphal fragment or clump of mycelium. Fragmentation of actinomycete mycelium does not occur readily in some genera and clumps are likely to be the predominant vegetative propagule. The use of molecular markers for detection and monitoring is therefore essential in ecological studies to overcome some of the above problems. An attempt was made to compare enumeration by culturing spores with molecular detection of plasmid markers genes in soil spores by probing soil DNA (15). The results indicated that viable counts underestimated the population by at least 10^2 for spores and for mycelial propagules the underestimate was greater than 10^3 c.f.u g^{-1}.

Molecular detection techniques are also essential in ecological studies for the discovery and monitoring of nonculturable bacteria which, for a range of reasons have eluded isolation. Detection of nonculturable pathogenic actinomycetes has been achieved by the amplification and sequencing of 16S rDNA from clinical specimens (16). It is possible that the soil contains other actinomycete groups still to be discovered due to exacting requirements for isolation and culture.

Tracking recombinant bacteria in soil also requires the use of definitive molecular monitoring and groups such as the *Streptomyces* genus may serve as hosts for expression of proteins (17). Strain improvement programmes for antibiotic-producing actinomycetes are also likely to result in the large scale use of recombinant strains.

11.2 Life-cycle of Streptomycetes in Soil

Studies of the culturable actinomycete populations in soil have indicated that *Streptomyces* species are the most abundant (18, 19). However, they appear to exist in soil mostly as spores and undergo short periods of comparatively rapid growth, which are discontinuous, during periods of nutrient availability (11, 20). Nutrients are in the form of particulate organic substrates such as fungal mycelium and fragments of roots which are colonized by mycelium (21). Total viable plate counts of soil extracts can only reflect variations in the relative abundance of streptomycete propagules, but spore-specific extraction methods allow estimation of the inactive component of viable counts, although vegetative mycelium can also be inactive but survive in soil (22). The development of molecular markers for detection and monitoring *in situ* has facilitated the study of growth and differentiation of *Streptomyces* species in soil by determining the presence of markers directly in soil DNA extracted from propagules in different stages of the life-cycle.

11.2.1 Spore Germination and Mycelial Development in Soil

Streptomycete spores are not highly resistant structures, being sensitive to heat but can survive desiccation and starvation conditions (23). The spore ribosome content is about only one third of that encountered in fully active mycelium. Spore ribosomes have lower activity (*i. e.* protein synthesis) than mycelial ribosomes, although this activity is quickly increased to the equivalent of fully active mycelium during the initiation of germination (24). This indicated the importance of these structures for dispersal and colonization, being in a state of readiness for germination if nutrients become available. Exogenous nutrients and water are essential for spore germination. Initiation of germination is not fully understood, but the presence of calcium ions appeared to be essential for this step in operating a calcium transport system causing excretion of dormancy-associated amino acids, ammonia and trehalose, resulting in ATP production (23). Trehalose may play an important role in the resistance of spores to moderate heating and desiccation (25).

Viable spore counting has been used effectively to monitor germination of *S. lividans* spores inoculated into sterile and nonsterile soil using the spore-specific extraction method of Herron and Wellington (26). This method involved differential isolation of streptomycete spores from soil samples by differential centrifugation and filtration. The mycelium was mostly entrapped in soil particles sedimented at a low centrifugation speed, whilst spores and suspended mycelial propagules were retained in the supernatant. This fraction was finally separated by high speed centrifugation as spores are more dense than mycelium. The resulting spore pellet generally contained less than 1% of the mycelia of the original sample (27). The method involved the use of the ion exchange resin Chelex-100 (Bio-Rad) which breaks up the soil structure by exchanging sodium ions from the eluent for calcium ions on the clay surfaces during extraction. This breaks down electrostatic bonds and disperses aggregates thus releasing spores from soil particles.

The life-cycle of *Streptomyces* species inoculated into soil has been observed by recording drops in viable spore counts which correspond to germination, and was followed by sharp rises, corresponding to re-sporulation. This cycle usually takes 5 days in sterile soil (26, 28, 29, 30) but only two days in nonsterile soil if nutrients have not been added (26, 28).

Once germ tubes emerge spores become trapped within the soil and lose density. Studies of life-cycles in nonsterile soil using the spore-specific method are, however, less sensitive due to low frequencies of germination of inoculated spores, and growth of indigenous microorganisms on plates inhibits accurate counting. Streptomycete spores on coverslips underwent 35% germination in sterile soil as opposed to 5% in natural soil (20). Once germinated however, there appeared to be little difference in the density of mycelial growth between sterile and nonsterile soil, the mean generation time in natural soil being 1.7 days (20).

The temperate actinophage KC301 (31), a derivative of Φ C31, has been used in the study of *S. lividans* life-cycle in soil. Peaks in soil phage titre corresponded to periods of host susceptibility to phage infection during germination and mycelial growth. KC301 carries the thiostrepton resistance marker gene (*tsr*), so lysogenic infections of both inoculated *S. lividans* strains (26, 28, 29, 30) and indigenous streptomycetes (29) have been detected by selection for this marker. KC301 has been used to enhance observations of the life-cycle of streptomycetes in soil, and have allowed precise determination of germination times. This was achieved by specifically extracting actinophages from soil inoculated with lysogenic host spores. It was possible to estimate the KC301 proportion released from lysogenic spores on germination by picking plaques onto lawns of *S. lividans*, transferring plaque DNA to nylon filters and probing these with labelled KC301 DNA. The results are illustrated in Fig. 1, which shows the release of KC301 into nonsterile soil (amended with chitin) corresponding to germination of spores. The burst in free KC301 represented about 1% of the lysogenic spores being induced.

11.2.2 Molecular Monitoring of Differentiation in Soil

The cell wall of streptomycete spores differs from that of the mycelium in its resilience to physical and chemical disruption. This difference has been exploited to ex-

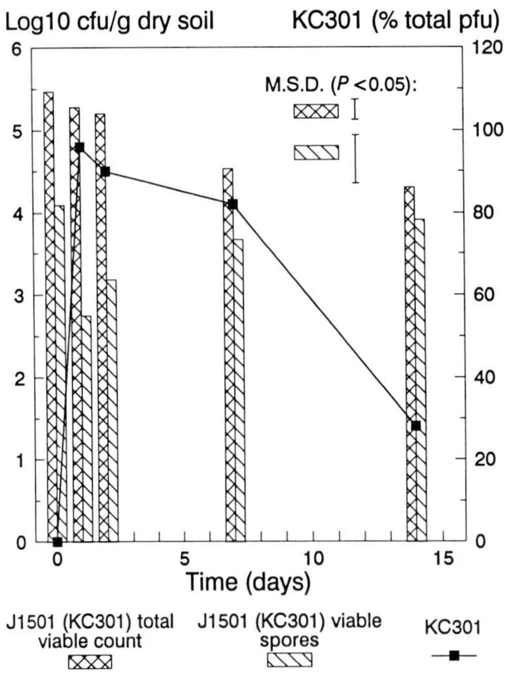

Fig. 1. Release of Phage KC301 from Lysogenic *Streptomyces coelicolor* Indicating the Time of Spore Germination

tract plasmid DNA differentially from spores and mycelia of streptomycetes inoculated into soil. SDS resistance of spores was noted by (32) who used it for selective inhibition of eubacteria enabling isolation of actinomycete spores from soil. Cresswell *et al* (33) used SDS/heat lysis treatment to extract DNA from the mycelial portion of *S. violaceolatus* ISP5438 (harbouring a stable marker plasmid pIJ673) in sterile soil samples. The spores were resistant to this treatment, therefore no DNA was extracted from them. Levels of plasmid pIJ673 were detected and quantified by probing Southern blots with radio-labelled pIJ673. Bead-beating (physical breakdown of the spore and mycelia walls) of soil samples yielded DNA from both spores and mycelia, and signals from Southern blots of extracts obtained by bead-beating were compared to those obtained by SDS/heat lysis. Thus by the use of differential DNA extraction methods and plasmid probes, the life-cycle of *S. violaceolatus* ISP5438 in soil could be studied. This technique was used to show that in a dynamic nonsterile soil microcosm system inoculated with *S. violaceolatus* ISP5438 (pIJ673), the inoculant existed mainly as spores after 17 days despite periodic nutrient amendment during the subsequent 53 days of incubation (28). Southern blot analysis only showed detection of mycelial DNA by SDS/heat lysis at days 15 and 17, whereas bead-beating resulted in detection throughout the incubation period.

Differential DNA extraction was used to demonstrate that in sterile soil of relatively low matric potential, mycelial growth and ensuing sporulation of *S. violaceolatus* ISP5438 (pIJ673) was much greater than at a higher matric potential, a conclusion that was not possible based on viable plate counts alone (34). Thus at 10% (w/w) moisture content (equivalent to a matric potential of −300 kPa), there was four times more DNA detected than at 20% (w/w) moisture content (−23 kPa) after 10 days of incubation in soil. The viable plate counts for spores and total propagules at day 10 were not significantly different. Soil moisture is known to influence streptomycete growth, which is most abundant in soils with pores which are humid and air filled as opposed to water-logged and less well aerated soils (34, 35).

Problems may occur in the use of molecular markers for the detection and monitoring of propagules in soil if there is a natural background to the marker. For monitoring streptomycete populations the antibiotic genes *tsr*, thiostrepton resistance, and *nptII*, neomycin resistance have proved useful as no indigenous background genes have been detected by probing both soil DNA and by PCR. The *xylE* gene proved to be a useful marker for streptomycetes in soil, but background signals were detected in some soils (36).

It is important to establish the presence of detectable introduced signal in cell-free DNA. This can be achieved by the omission of the lysis step in the DNA extraction leaving a simple alkali extraction. This method was used to detect the presence of extracellular DNA in samples of nonsterile soil inoculated with *S. lividans* TK64541Δ 2 (pIJ673). This strain had a reduced fitness resulting from amplification and deletion of DNA. It was unable to sporulate and was auxotrophic for proline and arginine. The survival of this disabled strain was studied in natural soil microcosms. Soil was seeded with 4.0 x 10^8 c.f.u. g^{-1} of mycelial propagules (homogenized) and incubated for 15 days. DNA extractions were made using SDS/lysis, bead-beating and a simple soil washing extraction which involved flushing soil samples with phosphate buffer (0.12 M; pH 8.0), and making standard DNA preparations from the supernatant (*i. e.* the SDS/heat lysis method minus the lysis step). The three types of DNA extracts were Southern blotted and probed with ^{32}P-labelled pIJ673 (Fig. 2). A signal was detected throughout the 15 days using the SDS/heat lysis and bead-beating methods, whereas extracellular pIJ673 DNA in soil was only detected at day 0 using the non-lysis washing method. This result suggested that DNA was released immediately after inoculation but became undetectable from day 1 onwards, probably due to degradation and binding of released DNA to soil aggregates. Turnover of free DNA released from streptomycetes in soil was therefore probably very rapid. SDS/heat lysis gave very smeared bands of DNA due to the presence of humic materials in the extract. The bead-beating method gave more defined bands indicating much less contamination of extracts. This illustrated that the much shorter extraction time of the bead-beating method prevented a build-up of humic materials in the extracts which interfered with Southern blot analysis. The presence of a detergent in the SDS/heat lysis method also caused increased extraction of phenolic components. Molecular techniques allowed detection of disabled strains which became unculturable on selective media (see section 11.4.2)

11.3 Potential for Genetic Interactions between Actinomycetes in Soil

The study of genetic interactions in natural ecosystems has been the focus of much interest over recent years largely because of concern over the release of genetically engineered microorganisms into the environment and spread and effects of recombinant DNA (reviews: 37, 38, 39). The nature of gene transfer mechanisms has been

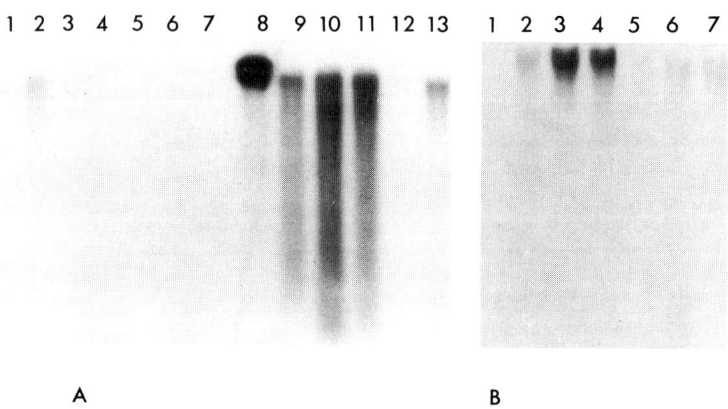

Fig. 2. Monitoring the fate of a disabled streptomycete in soil using three methods for DNA extraction. *Streptomyces lividans* TK64541Δ2 marked with pIJ673 marker plasmid. A, Lane 1, extracellular DNA, uninoculated control; lanes 2–7, days 0, 1, 2, 5, 10, 15. Lane 8, pIJ673; 9–13, SDS/lysis extracts, days 0, 1, 2, 5, 10. B, bead-beating extracts, lanes 1–7 as for A. D. McDowell, A.Vionis and E. M Wellington, unpublished data.

studied using laboratory strains in model environments and has demonstrated how genetic interactions possibly occur in natural microbial communities (38, 39). Hyphal fusion between actinomycetes facilitates gene exchange between different strains, and therefore conjugative transfer of plasmids may occur in this way, analogous to transfer of plasmid DNA via sex pili of Gram-negative bacteria (40). There is obviously potential for chromosomal and plasmid transfer to take place between streptomycetes as about 30% of wild-type *Streptomyces* species examined carry CCC (covalently closed circular) DNA plasmids which transfer very efficiently, and allow moderately efficient transfer of chromosomal genes (41). An important feature of certain streptomycete plasmids is their ability to spread through mycelium in addition to transfer abilities. Keiser (42) recognised a specific spread region on pIJ101 responsible for the movement of that plasmid in the mycelium. Some *Streptomyces* species plasmids such as pIJ702 will replicate in other actinomycete genera, for example *Micromonospora, Amycolatopsis, Saccharopolyspora* and *Thermomonospora* species.

Insertion sequences such as IS*117* found in *S. lividans* may be factors in causing gene exchange in streptomycetes, although there is no evidence at present for insertion within genes (41). Transposons have been detected in *Streptomyces* species such as Tn*4556*, the Tn*3*-like transposon from *Streptomyces fradiae*, and may be involved in gene rearrangement (40, 43).

Protoplast fusion *in vitro* has been used to recombine genes in crosses between actinomycetes such as *Nocardia asteroides* strains, although the occurrence of this type of transfer between actinomycetes in soil has not been investigated. Transformation in soil has not been considered either as actinomycetes are not known to be naturally competent.

There are at least two known transducing actinophages, SV1 and SH10 (44, 45), and although transduction has been demonstrated *in vitro*, it has yet to be observed in soil. Interspecific transfer of melanin genes on cosmids between *Streptomyces* species by a temperate actinophage has been shown (46), and underlines the possibilities for transduction in natural environments.

Conjugation and lysogenic conversion but not transduction have so far proved to be the modes of genetic interactions between streptomycetes in soil (26, 28, 29, 30, 47, 48, 49, 50). Gene transfer in soil may be a rare event because of the low levels of activity and spatial and temporal limitations to cell-to-cell contact.

11.3.1 Conjugative Interactions between Streptomycetes in Soil

Plasmid transfer via conjugation between streptomycetes requires hyphal fusion to take place although very little is known about the mechanism of transfer. Gene transfer by conjugative plasmids and mobilization of a nonconjugative plasmid was demonstrated in sterile soil (47). The donor used was *S. lividans* and recipient streptomycetes were inoculated into the soil, triparental crosses occurred with two donors, one containing either pIJ101 or pIJ211 (conjugative plasmids) and the other containing the nonconjugative plasmid pIJ702. This illustrates the possibilities for transfer from introduced streptomycetes containing nonconjugative plasmids to indigenous streptomycetes in soil where populations will contain conjugative plasmids (48). Wellington et al. (50) reported interspecific transfer of the plasmid pIJ673 between *S. violaceolatus* ISP5438 (pIJ673) and *S. lividans* TK24 (which contains no known plasmids) in nonsterile soil. Use of SEM observations showed that mycelial growth and therefore conjugation had occurred within two days of spore germination in soil, after which time re-sporulation had taken place and no further conjugative transfers were detected. Frequency of plasmid transfer was increased dramatically when the soil was nutrient amended, reflecting a correlation between metabolic activity *in situ* and genetic interactions. Peak transconjugant numbers and plasmid DNA in soil correlated with peak abundance of parental strains in the mycelial form, as observed in a study of conjugative transfer between *S. violaceolatus* ISP5438 (pIJ673) and *S. lividans* TK24 (28). A dynamic nonsterile soil microcosm system was used in this study to follow interactions where periodic soil replacement resulted in nutritional and spatial redistribution. In parallel experiments using the phage KC301 a correlation was found in the form of detection of lysogenic conversion of *S. lividans* TK24 at the time of peak transconjugant recovery. This enabled the precise monitoring of colonial development where peak mycelial development was pronounced.

11.3.2 Gene Exchange between Actinomycetes and Other Bacteria

Amongst the actinomycete genera, *Streptomyces* species have proved capable of expressing DNA cloned from a wide range of bacterial groups. More recent studies have indicated that cell to cell interactions can occur between Gram-negative bacteria and actinomycetes resulting in transfer of DNA. IncQ plasmids can replicate in a range of bacterial hosts. The plasmid RSF1010 belongs to this class, and although it is nonconjugative, it can be mobilized between different Gram-negative species. Gormley and Davies (51) showed that *E. coli* could transfer RSF1010 to *S. lividans* and *Mycobacterium smegmatis*, in which the plasmid was stably maintained through seven generations whilst conferring its antibiotic resistances (streptomycin and sulfonamide) to the recipients. Mazodier *et al.* (52) also noted the ability of *E. coli* to transfer specially constructed shuttle plasmids to various *Streptomyces* species by conjugation. The plasmids contained origins of replication from pBR322 and pIJ101 and the RK2 (IncP) origin of transfer.

Natarajan and Oriel (53) demonstrated transposition involving actinomycetes in soil using *Bacillus subtilis* strain containing Tn*916*. Indigenous actinomycetes belonging to the genera *Streptomyces*, *Actinomadura* and *Kineosporia* acquired the transposon and were detected by tetracycline resistance and confirmed by probing. This was the first report of an actinomycete acquiring Tn*916*.

The dissemination of antibiotic resistance genes from producing organisms such as streptomycetes could have occurred by conjugal transfer events although experimentally it has only been possible to detect transfers by conjugation with Gram-negative bacteria involving acquisition of genes by actinomycetes. The presence of related genes in diverse bacterial groups, such as similar streptomycin phosphotransferase genes in *S. griseus* and *Mycobacterium fortuitum*, or *Staphylococcus* species and Gram-negative R-plasmids, suggests a movement of genes from certain common sources throughout the history of bacterial evolution (54).

11.3.3 Interactions between Streptomycetes and Actinophages in Soil

Bacteriophages active against actinomycetes are readily isolated from soil, and those active against streptomycetes are the most common actinophages isolated (55). Homologous phage modules (functional sets of genes) have been found in unrelated temperate and virulent actinophages of *Faenia* and *Saccharopolyspora* species (56). The presence of this common DNA in phages which infect different actinomycete genera suggests a potential for phage-mediated interactions between different genera. Due to the spasmodic growth pattern of streptomycetes in soil microsites, opportunities for infection by actinophages are limited to when spores germinate and mycelial growth takes place. Virulent phages must be able to persist in soil during long periods of host inactivity (57). Actinophage will bind non-specifically to resting streptomycete spores (58), although actual infective penetration of streptomycetes by the phage genome is only possible via germ tubes and newly growing hyphal tips, spores are resistant to phage infection (59).

Phases of host growth in soil correlate with large increases in phage titre in sterile and nonsterile soils inoculated with *S. lividans* and the genetically engineered temperate actinophage KC301 (28, 29, 30). KC301 survives poorly in the free state in nonsterile soil compared to indigenous actinophages (28). Selection for *tsr* carried on the phage genome allowed monitoring of host conversion in soil by lysogeny. Indigenous streptomycetes in nonsterile soil were lysogenized by KC301 and could be detected by isolating thiostrepton resistant indigenous streptomycete spores, which proved on molecular analysis to contain integrated KC301 DNA in the host chromosome (29]. An engineered actinophage was thus shown to be capable of surviving in nonsterile soil as a prophage within streptomycete spores. KC301 was stably integrated into the chromosome via the phage *att* site, and on germination expressed the phenotype carried on the prophage so lysogenic conversion had occurred. The presence of the KC301 prophage reduced the fitness of lysogenized *S. lividans* hosts in sterile soil. This was manifested as reduced maximum population densities compared to their uninfected counterparts (30). Germinating lysogen spores released KC301 into sterile soil which lysogenically infected susceptible hosts if present. Hence intraspecific gene transfer mediated by KC301 (lysogenic conversion) occurred in sterile soil between different *S. lividans* strains (30). Interspecific gene transfer using lysogenized *S. coelicolor* and uninfected *S. lividans* in sterile soil was also demonstrated.

11.4 Detection and Expression of Specific Genes in Soil

The production of antibiotics by many streptomycetes, all of which were isolated from soil or plant parts, has prompted much research and speculation into the possible production and role of antimicrobial secondary metabolites in soil (60). Although antibiotics have not been isolated from natural untreated soils, production has been detected in nonsterile amended and sterile unamended and amended soils when inoculated with producing strains of streptomycetes and fungi (61, 62). There is evidence to support the occurrence of antibiosis in soil, for example survival of a *Salmonella* species in soil was inhibited by the presence of streptomycin-producing *Streptomyces bikiniensis* (63). In another example *S. hygroscopicus* var. *geldanus* (a geldanamycin producer) was grown in sterile soil 2 days prior to infestation of the soil with the pea root rot fungus *Rhizoctonia solani* and planting with its host. The disease was controlled and extracts from the soil were found to contain geldanamycin and inhibited *R. solani* growth (10). The antibiotic was present in the soil at 88 μg/g after growth of the producer.

The occurrence of antibiosis has been tested in a simple chemostat system where *S. aureofaciens* (a tetracycline producer) was grown together with sensitive *E. coli* or *Bacillus pumillus* strains (64). When tetracycline was produced in the culture, the sensitive strains were completely inhibited. There is a growing body of evidence that actinomycetes and other antibiotic-producing bacteria may be antagonistic to the growth of sensitive strains in soil, particularly when growth of producers is in close proximity to plants such as in the rhizosphere where a high nutrient status leads to detectable levels of secondary metabolites (7, 8, 10, 65). Streptomycete strains that were found to be active colonizers of potato roots and tubers under low matric potential were also found to be bioactive, and several of these strains were streptomycin producers (3).

It is possible that streptomycete secondary metabolites play a role in cellular differentiation and physiology (66, 67). Thiostrepton induces the production of several proteins via its effect on a thiostrepton sensitive promoter (*tipA*) in *S. lividans* (68). The role and function of these proteins is currently unknown, but the presence of *tipA* in a sensitive non-producing streptomycete does indicate prior exposure to this antibiotic in the soil environment. Gene probes based on genes common to the production of a wide range of streptomycete natural products may be used to detect the potential for secondary metabolite production in a broad range of isolates (69, 70). Evidence for antibiotic production in soil is currently achieved by detection of active product extracted from the soil. There may be a potential for detecting activity of specific promoters associated with antibiotic production in soil samples by the use of reporter genes such as *lux* operon from the luminescent marine bacteria *Vibrio fischeri* and *V. harveyi*. Expression *in situ* of an operon required for antibiotic biosynthesis was detected using a *lux* gene transcriptional fusion. Gene expression was monitored by measuring β-galactosidase activity in strains of *Pseudomonas fluorescens* grown on cotton seed (9). *Lux* reporter gene transcriptional fusions have been used in streptomycetes to study differentiation (71). Light production in different parts of the colony indicated temporal and spatial control of developmental genes. This type of *luxAB* fusion could also be used to indicate potential expression of antibiotic production in soil, although currently light production levels with these genes are very low.

11.4.1 Antibiotic Resistance Genes and Expression of Antibiotic Production Genes in Soil

Antibiotic resistance genes have proved useful as molecular markers for studying the streptomycete life cycle in soil (see section 11.2). The presence of indigenous antibiotic resistance and production genes in soil populations can be investigated directly by probing soil DNA and use of specific PCR primers. DNA homologous to *aphD*, required for streptomycin production, was detected in DNA extracted from a soil which had undergone liming (63). This indicated the presence of a population of actinomycetes with the potential for antibiotic production, probably in excess of 10^3 c.f.u. g^{-1}, which is the detection limit for DNA from soil samples using PCR. The effect of the liming of this acid soil was to increase the bioactive population of streptomycetes in the soybean rhizosphere. This resulted in competition and antagonism with inoculated *Bradyrhizobium* strains.

Many antibiotics are produced towards the end of vegetative growth at the onset of sporulation, and therefore it may be hypothesized that these compounds act as an extracellular protection mechanism for dormant spores in soil, inhibiting invasion of a microsite by other bacteria. Thiostrepton is a polypeptide antibiotic and a natural product of *S. azureus*, originally a soil isolate. When added to soil this antibiotic caused death of the mycelial stage of sensitive *S. lividans*, whereas the insertion of the thiostrepton resistance gene *tsr* (via lysogenic conversion by KC301) created a resistant strain which was unharmed by the antibiotic in soil (30). Interestingly, sporulation compensated for killing of the mycelial stage as spores are resistant, allowing propagule numbers to reach a level equivalent to sensitive strains grown in the absence of thiostrepton. Additionally, killing of the mycelial stage due to lysis by actinophages in soil was similarly compensated for by sporulation (30).

Thiostrepton can be extracted from soil samples using ethyl acetate, followed by quantification by bioassaying redissolved extracts (in chloroform) against supersensitive bacteria or by use of a *tipA* promoter gene fusion (68). Use of the extraction and assay techniques showed that thiostrepton was produced following sporulation of *S. azureus* spores in soil and was detected at a level of about 0.05 $\mu g.\ g^{-1}$ (63). Furthermore, nutrient amendments of soil resulted in increased mycelial activity (*i. e.* germination) of *S. azureus*, which corresponded to increased thiostrepton activity in soil (Fig. 3). Poor nutrient availability in natural soil may result in very low levels of antibiotic being produced, hence problems in detection. However, from sterile soil experiments, it is clear that in soil antibiotic production correlated with differentiation and sporulation.

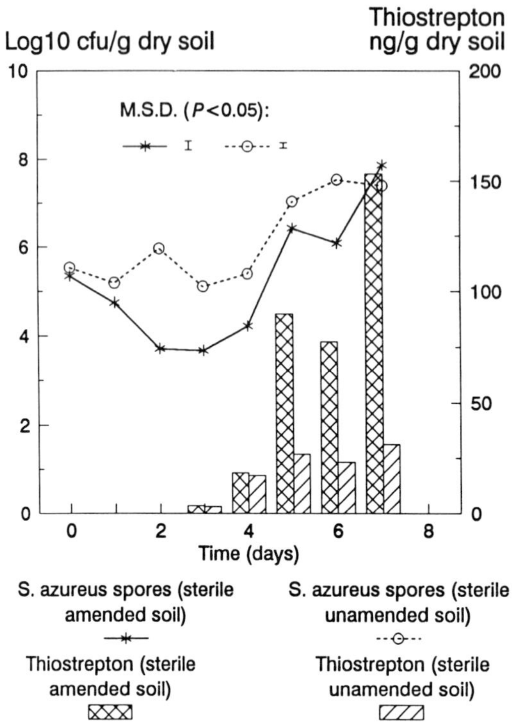

Fig. 3. Thiostrepton production in soil by *Streptomyces azureus* and its correlation with sporulation. P. Marsh, M. Meijer and E. M Wellington, unpublished data.

11.4.2 Detection of Amplified Genes in Soil

An unusual feature of *Streptomyces* species is their ability to spontaneously amplify specific chromosomal DNA sequences in the absence of selection, with a corresponding deletion. This property has been exploited for detection in soil using cloned resistance genes by linking them to the amplified sequence. Amplified DNA containing marker genes was introduced by transformation into *S. lividans* 66 (72). This property was exploited to detect the agarase gene *dag* originally from *S. coelicolor* (73), which had been introduced, linked to the amplified sequence, into *S. lividans* TK64 541Δ2 (Altenbuchner, unpublished data). DNA from soil extracts (sterile unamended soil inoculated with 10^6 c.f.u. g^{-1} of *S. lividans* TK64 541Δ2 and sampled on days 2 and 5, lanes 2 and 3 in Fig. 4A, B) and preparations from broth cultures were Southern blotted and probed with ^{32}P-labelled *dag* (Fig. 4). Viable counts were about 6 x 10^5 c.f.u. g^{-1} on days 2 and 5. Amplification facilitated detection of this gene in soil, and genes such as *dag* are useful in soil experiments as neither *S. coelicolor* nor *dag* have so far not been detected in any soil.

In addition the *dag* phenotype resulted in degradation of agar producing characteristic depressions around the colonies. Such depressions were observed even before

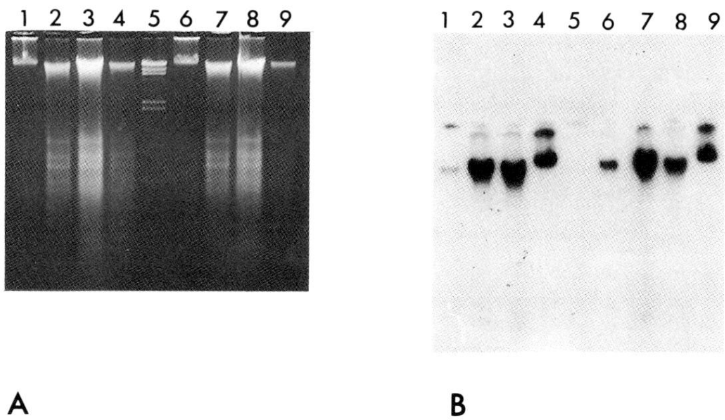

Fig. 4. Detection of the marker gene for agarase production, *dag,* in sterile soil inoculated with *Streptomyces lividans* TK64 541Δ2 carrying amplified copies. A, agarose gel; B, Southern blot using a *dag* probe; Lane 1, uninoculated soil; lane 2, soil day 2; lane 3, soil day 5; lanes 4–9, DNA from mycelium grown in shake flasks. D. McDowell, A.Vionis and E. M Wellington, unpublished data

the colonies were seen by the naked eye. This marker gene thus allowed both genotypic and phenotypic detection of an inoculant actinomycete. Detection limits were investigated for the *dag* marker and found to be in the region of 10^4 c.f.u. g^{-1} without PCR amplification of the DNA. In nonsterile soil the disabled strain TK64 541Δ2 became difficult to culture using the selective media containing thiostrepton and neomycin. Under these conditions it was still possible to monitor growth and survival using the *dag* or *tsr* genotypes.

S. lividans DNA extracted from soil appeared to be digested on occasions as seen in Fig. 4A, lane 5. This was probably due to site-specific degradation of the DNA during electrophoresis in buffer contaminated with ferrous iron (74).

11.5 Conclusions

The use of molecular monitoring has greatly facilitated ecological studies of actinomycetes in soil and allowed detection of populations containing specific genes. The monitoring directly of gene expression in soil is still problematic although some limited studies have been done using transcriptional fusions to reporter genes. *In situ* hybridizations using 16S rRNA probes targeting ribosomes did indicate active mycelium of *Streptomyces scabies* in soil as the fluorescence was related to ribosome number (3). A range of marker genes in addition to 16s rRNA have been used to detect populations in soil and the rhizosphere but relating molecular data to traditional plate counts has proved difficult in some cases. Actinomycetes are thought to be readily culturable but still many problems exist in accurate quantification of a mycelial bacterium in soil. Profuse sporulation may result in high viable counts which only represent inactive populations. Quantitative PCR may enable more accurate estimation of marked populations and this is possible for indigenous and introduced molecular markers.

Acknowledgements

We gratefully acknowledge financial support from the EC BRIDGE programme (BIOT 910285) and Natural Environment Research Council (PM studentship).

11.7 References

1. McCarthy, A. J. Lignocellulose-degrading actinomycetes. FEMS Microbiol. Rev. 46, (1987) 145–163.
2. Crawford, D. L. Biodegradation of agricultural and urban wastes, In: Actinomycetes in Biotechnology (Goodfellow, M., Williams, S.T., and Mordarski, M. Eds.), (1988) pp. 433–439. Academic Press, London.
3. Bramwell, P. The characterisation and detection of plant pathogenic streptomycetes in the natural environment. Ph.D. Thesis, (1992) University of Warwick. U.K.
4. Watson, E.T. and Williams, S.T. Studies on the ecology of actinomycetes in soil – VII. Actinomycetes in a coastal sand belt. Soil Biol. Biochem. 6, (1974) 43–52.
5. Milus, E.A. and Rothrock, C.S. Rhizosphere colonization of wheat by selected soil bacteria over diverse environments. Can. J. Microbiol. 39, (1993) 335–341.
6. Elliott-Juhnke, M., Mathre, D.E. and Sands, D.C. Identification and characterization of rhizosphere-competent bacteria of wheat. Appl. Environ. Microbiol. 53, (1987) 2793–2799.
7. Thomashow, L.S., Weller, D.M., Bonsall, R.F. and Pierson III, L.S. Production of the antibiotic phenazine-l-carboxylic acid by fluorescent *Pseudomonas* species in the rhizosphere of wheat. App. Environ. Microbiol. 56, (1990) 908–912.
8. Weller, D.M. and Thomashow, L.S. Antibiotics: Evidence for their production and sites where they are produced, In: New Directions in Biological Control: Alternatives for Suppressing Agricultural Pests and Diseases (Baker, R.R. and Dunn, P.E. Eds.), (1990) pp. 703–711. Alan R. Liss Inc., New York.
9. Howie, W.J. and Suslow, T.V. Role of antibiotic biosynthesis in the inhibition of *Pythium ultimum* in the cotton spermosphere and rhizosphere by *Pseudomonas fluorescens*. Mol. Plant-Microbes Interact. 4, (1991) 393–399.
10. Rothrock, C.S. and Gottlieb, D. Role of antibiosis of *Streptomyces hygroscopicus* var. *geldanus* to *Rhizoctonia solani* in soil. Can. J. Microbiol. 30, (1984) 1440–1447.
11. Williams, S.T. Streptomycetes in the soil ecosystem, In: Nocardia and Streptomyces (Mordarski, M., Kurytowicz, W. and Jeljaszewicz, J., Eds.), (1978) pp. 137–144. Gustav Fischer Verlag, Stuttgart.
12. Ramos, M.L.G., Magalhaes, N.F.M. and Boddy, R.M. Native and inoculated rhizobia isolated from field grown *Phaseolus vulgaris*: effects of liming an acid soil on antibiotic resistance. Soil Biol. Biochem. 19, (1987) 179–185.
13. Roughley, R.J., Wahab, F.A. and Sundram, J. Intrinsic resistance to streptomycin and spectinomycin among root-nodule bacteria from malaysian soil. Soil Biol. Biochem. 24, (1992) 715–716.
14. Wellington, E.M.H. and Toth, I.K. Actinomycetes In: Methods of soil analysis, part 2 Chemical and Microbiological Properties -Agronomy Monograph no. 9 (3rd edition) ASA-SSSA. In press.
15. Wellington, E.M.H., Cresswell, N. and Herron, P.R. (1992) Gene transfer between streptomycetes in soil. Gene, 115, (1992) 193–198.
16. Relman, D.A., Thomas, M.D., Schmidt, T.M., Richard, P., MacDermott, M.D. and Falkow, S. Identification of the uncultured bacillus of whipple's disease. New Eng. J. Med., 327, (1992) 293–301.
17. Taguchi, S., Kumagai, I., Nakajama, J., Suzuki, A. and Miura, K. Efficient extracellular expression of a foreign protein in *Streptomyces* using secretory protease inhibitor (SSI) gene fusions. Biotechnol. 7, (1989)1063.
18. Locci, R. Streptomycetes and related genera, In: Bergey's Manual of Systematic Bacteriology, vol. 4, 9th edn. (Williams, S.T., Sharpe, M.E. and Holt, J.G., Eds.), (1989) pp. 2451–2508. Williams and Wilkins, Baltimore.
19. McCarthy, A.J. and Williams, S.T. Methods for studying the ecology of actinomycetes. Meth. Microbiol. 22, (1990) 533–563.

20. Mayfield, C.I., Williams, S.T., Ruddick, S.M. and Hatfield, H.L. Studies on the ecology of actinomycetes in soil IV. Observations on the form and growth of streptomycetes in soil. Soil Biol. Biochem. 4, (1972) 79-91.
21. Goodfellow, M. and Simpson, K.E. Ecology of streptomycetes. Front. Appl. Microbiol. II, (1985) 97-125.
22. Baker, P.W. The survival of *Streptomyces* species and mutant derivatives in soil. M.Sc. Thesis, (1993) University of Warwick, U.K.
23. Ensign, J.C., McBride, M.J., Stoxen, L.J., Bertinusen, A., Pomplun, M. and Ho, A. The life cycle of *Streptomyces*: germination and properties of spores and regulation of sporulation, In: Biological, Biochemical and Biomedical Aspects of Actinomycetes (Szabo, G., Biro, S. and Goodfellow, M., Eds.), (1985) pp. 777-790. Akademiai Kiado, Budapest.
24. Quiros, L.M., Parra, F., Hardisson, C. and Salas, J.A. Structural and functional analysis of ribosomal subunits from vegetative mycelium and spores of *Streptomyces antibioticus*. J. Gen. Microbiol. 135, (1989) 1661-1670.
25. McBride, M.J. and Ensign, J.C. Effects of intracellular trehalose content on *Streptomyces griseus* spores. J. Bacteriol. 169, (1987) 4995-5001.
26. Herron, P.R. and Wellington, E.M.H. New method for the extraction of streptomycete spores from soil and application to the study of lysogeny in sterile amended and nonsterile soil. Appl. Environ. Microbiol. 56, (1990) 1406-1412.
27. Herron, P.R. and Wellington, E.M.H. Extraction of *Streptomyces* spores from soil and detection of rare gene transfer events, In: Genetic Interactions Among Microorganisms in the Natural Environment (Wellington, E.M.H. and van Elsas, J.D., Eds.), (1992) pp. 91-103. Permagon Press, Oxford.
28. Cresswell, N., Herron, P.R., Saunders, V.A. and Wellington, E.M.H. The fate of introduced streptomycetes, plasmid and phage populations in a dynamic soil system. J. Gen. Microbiol. 138, (1992) 659-666.
29. Marsh, P. and Wellington, E.M.H. Interactions between actinophage and their streptomycete hosts in soil and the fate of phage borne genes, In: Gene transfers and environment (Gauthier, M.J., Ed.), (1992) pp. 135-142. Springer-Verlag, Berlin.
30. Marsh, P., Toth, I.K., Meijer, M., Schilhabel, M.B. and Wellington, E.M.H. (1993) Survival of the temperate phage Φ C31 and *Streptomyces lividans* in soil and the effects of competition and selection on lysogens. FEMS Microbiol. Ecol. 13, (1993) 13-22.
31. Chater, K.F. Bruton, C.J. King, A.A. and Suarez, J.E. The expression of *Streptomyces* and *Escherichia coli* drug-resistance determinants cloned into the *Streptomyces* phage C31. Gene, 19, (1982) 21-32.
32. Nonomura, H. and Hayakawa, M. New methods for the selective isolation of soil actinomycetes, In: Biology of Actinomycetes ,88 (Okami, Y., Beppu, T. and Ogawara, H., Eds.), (1988) pp. 288-293. Japan Scientific Societies Press, Tokyo.
33. Cresswell, N. Saunders, V.A. and Wellington, E.M.H. Detection and quantification of *Streptomyces violaceolatus* plasmid DNA in soil. Lett. Appl. Microbiol. 13, (1991) 193-197.
34. Karagouni, A.D. Vionis, A.P. Baker, P.W. and Wellington, E.M.H. The effect of soil moisture content on spore germination, mycelium development and survival of a seeded streptomycete in soil. Microbial Rel. 2, (1993) 47-51.
35. Williams, S.T., Shameemullah, M., Watson, E.T. and Mayfield, C.I. Studies on the ecology of actinomycetes in soil VI. The influence of moisture tension on growth and survival. Soil Biol. Biochem. 4, (1972) 215-255.
36. Wipat, A., Wellington, E.M.H. and Saunders, V.A. *Streptomyces* marker plasmids for monitoring survival and spread of Streptomycetes in soil. Appl. Environ. Microbiol. 57, (1991) 3322-3330.
37. Stotzky, G. and Babich, H. Survival of, and genetic transfer by, genetically engineered bacteria in natural environments. Adv. Appl. Microbiol. 31, (1986) 93-138.
38. Trevors, J.T., Barkay, T. and Bourquin, A.W. Gene transfer among bacteria in soil and aquatic environments: a review. Can. J. Microbiol. 33, (1987) 191-198.
39. Stotzky, G. Devanas, M.A. and Zeph, L.R. Methods for studying bacterial gene transfer in soil by conjugation and transduction. Adv. Appl. Microbiol. 35, (1990) 57-169.
40. Hopwood, D.A., Kieser, T., Lydiate, D.J. and Bibb, M.J. *Streptomyces* plasmids: their biology and use as cloning vectors, In: The Bacteria, vol. IX, The Antibiotic Producing Streptomycetes (Queener, S.W. and Day, L.E., Eds.) (1986) pp. 159-229. Academic, New York.

41. Kieser, K. and Hopwood, D.A. Genetic manipulation of *Streptomyces*: integrating vectors and gene replacement. Meth. Enzymol. 204, (1991) 430–458.
42. Keiser, T., Hopwood, D.A., Wright, H.M. and Thompson, C.J. pIJ101, a multicopy-copy broad host-range plasmid: functional analysis and development of DNA cloning vectors, 185, (1982) 223–238.
43. Chater, K.F., Henderson, D.J., Bibb, M.J. and Hopwood, D.A. Genome flux in *Streptomyces coelicolor* and other streptomycetes and its possible relevance to the evolution of mobile antibiotic resistance determinants, In: Transposition, Society of General Microbiology Symposium 43 (Kingsman, A.J., Chater, K.F. and Kingsman, S.M., Eds.), (1988) pp. 7–42.
44. Stuttard, C. Transduction of auxotrophic markers in a chloramphenicol producing strain of *Streptomyces*. Can. J. Microbiol. 110, (1979) 479–482.
45. Suess, F. and Klaus, S. Transduction in *Streptomyces hygroscopicus* mediated by the temperate phage SH10. Mol. Gen. Genet. 181, (1981) 552–555.
46. Morino, T., Tagaki, K., Nakamura, T., Takita, T., Saito, H. and Takahashi, H. (1988) Interspecific transfer and expression of melanin gene(s) on cosmids in *Streptomyces* strains. appl. microbiol. Biotech. 27, (1988) 517–520.
47. Rafii, F. and Crawford, D.L. Transfer of conjugative plasmids and mobilization of a nonconjugative plasmid between *Streptomyces* strains on agar and in soil. Appl. Environ. Microbiol. 54, (1988) 1334–1340.
48. Rafii, F. and Crawford, D.L. Donor/recipient interactions affecting plasmid transfer among *Streptomyces* species: a conjugative plasmid will mobilize nontransferable plasmids in soil. Curr. Microbiol. 19, (1989)115–121.
49. Bleakley, B.H. and Crawford, D.L. The effects of varying moisture and nutrient levels on the transfer of a conjugative plasmid between *Streptomyces* species in soil. Can. J. Microbiol. 35, (1989) 544–549.
50. Wellington, E.M.H., Cresswell, N. and Saunders, V.A. Growth and survival of streptomycete inoculants and extent of plasmid transfer in sterile and nonsterile soil. Appl. Environ. Microbiol. 56, (1990) 1413–1419.
51. Gormley, E.P. and Davies, J. Transfer of plasmid RSF1010 by conjugation from *Escherichia coli* to *Streptomyces lividans* and *Mycobacterium smegmatis.* J. Bact. 173,(1991) 6705–6708.
52. Mazodier, P., Petter, R. and Thompson, C. Intergeneric conjugation between *Escherichia coli* and *Streptomyces* species. J. Bact. 171, (1989) 3583–3585.
53. Natarajan, M.R. and Oriel, P. Transfer of transposon Tn*916* from *Bacillus subtilis* into a natural soil population. Appl. Environ. Microbiol. 58, (1992) 2701–2701.
54. Davies, J. Another look at antibiotic resistance. J. Gen. Microbiol. 138, (1992) 1553–1559.
55. Williams, S.T. and Lanning, S. Studies of the ecology of streptomycete phage in soil, In: Biological, Biochemical and Biomedical aspects of Actinomycetes (Ortiz-Ortiz, L., Bojalil, L.F. and Yakoleff, V., Eds.) (1984) pp. 473–483. Academic Press, London.
56. Schneider, J. and Kutzner, H.J. Distribution of modules among the central regions of the genomes of several actinophages of *Faenia* and *Saccharopolyspora*. J. Gen. Microbiol., 135, (1989) 1671–1678.
57. Williams, S.T., Mortimer, A.M. and Manchester, L. The ecology of soil bacteriophage. In: Phage Ecology (Goyal, S.M., Gerba. C.P. and Bitton, G. Eds.) (1987) pp. 157–179. John Wiley, New York.
58. Novikova, N.L., Lomovskaya, N.D. and Kapitonova, O.N. Adsorption and development of f C31 actinophage in germinatinng spores of *Streptomyces coelicolor* A3(2). Mikrobiologiya, 42, (1973) 513–518.
59. Lomovskaya, N.D., Chater, K.F. and Mkrtumian, N.M. Genetics and molecular biology of *Streptomyces* bacteriophages. Microbiol. Rev., 44, (1980) 206–229.
60. Williams, S.T. and Vickers, J.C. The ecology of antibiotic production. Microbial Ecol. 12, (1986) 43–52.
61. Williams, S.T. Are antibiotics produced in soil? Pedobiologia, 23, (1982) 427–435.
62. Wellington, E.M.H., Marsh, P., Toth, I. Cresswell, N., Huddleston, L. and Schilhabel, M. . The selective effects of antibiotics in soils. In Trends in Microbial Ecology, (Guerrero, R. and Pedros-Alio, C., Eds.) (1993) pp 331–336, Spanish Society for Microbiology, Spain.
63. Turpin, P.E., Dhir, V.K. Maycroft, K.A., Rowlands, C. and Wellington, E.M.H. The effect of *Streptomyces* species on the survival of *Salmonella* in soil. FEMS Microbiol. Ecol., 101, (1992) 271–280.

64. Rasool, K.A. and Wimpenny, W.T. Mixed continuous culture experiments with an antibiotic-producing streptomycete and *Escherichia coli.* Microbial Ecol., 8, (1982) 267–277.
65. Weller, D.M. and Thomashow, L.S. Antibiotics: Evidence for their production and sites where they are produced. In New Directions in Biological Control: Alternatives for Suppressing Agricultural Pests and Diseases, (1990) pp. 703–711, (Baker, R.R. and Dunn, P.E., Eds.) Alan R. Liss, Inc., New York.
66. Murakami, T., Holt, T. and Thompson, C.J. The regulation of a thiostrepton induced promoter in *Streptomyces lividans.* In Biology of actinomycetes ,88, Okami, Y. Beppu T. and Ogawa, H. Eds.) (1988) pp. 371–373. Japan Scientific Societies Press, Tokyo.
67. Beppu, T. Secondary metabolites as chemical signals for cellular differentiation. Gene, 115, (1992) 159–165.
68. Murakami, T., Holt, T.J. and Thompson, C.J. Thiostrepton-induced gene expression in *Streptomyces lividans.* J. Bacteriol. 171, (1989) 1459–1466.
69. Phillips, L., Wellington, E.M.H., Rees, S.B. and King, G.P. The distribution of DNA sequences homologous to antibiotic production and resistance genes within type strains and wild isolates of *Streptomyces* species. J. Antibiotics 45, (1992) 1481–1491.
70. Stockmann, M. and Piepersberg, W. Gene probes for the detection of 6-deoxyhexose metabolism in secondary metabolite-producing streptomycetes. FEMS Microbiol. Lett., 90, (1992) 185–190.
71. Schauer, A., Ranes, M., Santamaria, R., Guijarro, J., Lawlor, E., Mendez, C., Chater, K. and Losick, R. Visualizing gene expression in time and space in the filamentous bacterium *Streptomyces coelicolor.* Science 240, (1988) 768–772.
72. Altenbuchner, J. and Cullum, J. Amplification of cloned genes in *Streptomyces.* Biotechnology, 5, (1987) 1328–1329.
73. Kendall, K. and Cullum, J. Cloning and expression of an extracellular-agarase gene from *S. coelicolor* A3(2) in *Streptomyces lividans* 66. Gene, 29,(1984) 315–321.
74. Zhou, X., Deng, Z., Firmin, J.L., Hopwood, D.A. and Kieser, T. Site specific degradation of *Streptomyces lividans* DNA during electrophoresis in buffers contaminated with ferrous iron. Nucleic Acids Res. 16, (1988) 4341–4352.

12 Some Considerations on Gene Transfer between Bacteria in Soil and Rhizosphere

J.D. van Elsas and E. Smit

12.1 Introduction

The worldwide dramatic increase in agricultural and industrial productivity has created severe environmental problems. For instance, soil and groundwater reservoirs have been polluted with pesticides, xenobiotics and heavy metals. Biological agents, such as bacteria with beneficial properties, could in some cases substitute pesticides (biological control), or serve to remediate contaminated environments like soil and sediment. Bacteria have already been introduced into soil to promote plant growth (1, 2, 3) for pest control (4, 5, 6) or for the degradation of a variety of polluting compounds (7, 8, 9). Although many bacteria are naturally capable of performing specific functions, the range of possibilities is limited, since introduced bacteria do not always survive and perform well in the soil ecosystem.

Bacteria isolated from the environment, which are potentially better adapted to ecological stresses, can be genetically altered for specific environmental purposes (10, 11, 12, 13, 14, 15, 16, 17). The use of recombinant microorganisms in large-scale field trials is, however, still restricted. This is due to a lack of knowledge on the fate of the microorganism and/or the heterologous DNA and its possible ecological and health effects as a consequence of the release (18, 19, 20, 21).

One area of particular concern is the transfer of introduced genes to indigenous microbes (21). Evidence which suggested that environmental transformation, transduction and conjugation take place under favourable conditions, has accumulated over the last 5 years (21, 22, 23). Data on gene spread were obtained both directly, by following transfer of an introduced genetic element from donor to recipient microorganisms, and retrospectively, by tracking similar genetic elements in different microorganisms in the environment.

Since many environmental releases of genetically engineered bacteria will be into soil, e.g. in biological control (16) in bioremediation (9), in forestry and in biomining, our knowledge on the conduciveness of soil to genetic interactions between introduced and indigenous bacteria should be enhanced. Soil is a dynamic environment with special characteristics for interactions between its bacterial inhabitants. Several chapters in recent books on environmental gene transfer (21, 22, 23) have addressed gene transfer in soil. Rather than duplicating these substantial contributions, this paper will evaluate the present status of knowledge on gene transfer processes in soil and rhizosphere, with regards to the intricacies of the soil environment.

12.2 Soil and Rhizosphere as Habitats for Bacteria

Soil represents a three-phase environment composed of solid, liquid and gaseous phases. The solid phase is static, as opposed to the liquid and gaseous phases, where conditions are dynamic and commonly fluctuate. All 3 phases are heterogeneous with regards to the distribution of different (gaseous, liquid or solid) compounds (24). The solid phase contains inorganic substances such as clay, silt, sand and organic matter (humic substances), which generally are distributed unevenly (25) and are often complexed in aggregates of varying sizes, composition and stability. Aggregates important for soil microorganisms are clay-organic matter complexes, due to their negatively charged surfaces and increased nutrient availability (24, 26). The solid phase is interspersed with the soil pore network containing liquid and gaseous phases of varying composition. The volume of pores with different sizes is governed by soil structure and texture and determines the moisture retention capacity of the system, and thereby the availability of water for soil microorganisms.

Bacteria in soil are located in the often discontinuous soil pores, and are commonly associated with soil surfaces. In the absence of transporting agents such as (multidirectional) water flow, growing plant roots or burrowing soil animals, bacterial movement in soil over large distances is limited (27). Bacteria may therefore be confined to soil sites where they are located. The conditions at the level of each individual soil site (pore), rather than at the overall soil level, determine the fate of the bacteria present. It is very difficult to study the bacteria and processes taking place at the individual pore level in soil (28). Consequently, bacteria in soil are commonly studied on a larger scale which may not provide information on individual bacterium-pore relationships.

Bulk soil can be regarded as an oligotrophic environment, since it is generally poor in readily-available organic carbon (29). For instance, even though a loamy sand soil studied in our laboratory contained 100 μg C per g., this carbon was suggested to be largely unavailable to microorganisms since it is recalcitrant or localized in sites inaccessible to soil bacteria (30). The resulting low amount of available carbon in soil generally precludes much bacterial growth and activity, and is estimated to be sufficient for only a few cellular divisions of soil microorganisms per year (31, 32). Nutrients may, however, become transiently available in localized "hot spots?", e. g. in decaying material of plant/animal origin or in the rhizosphere, and trigger microbial growth and activity.

Plant roots are major sites of input of carbon into soil (33, 34). Both water-soluble compounds and insoluble ones like remnants of root cortex cells, are released into the rhizosphere. For example, between 3 and 30 times more soluble organic matter could be present in the rhizosphere of gramineous plants than in the corresponding bulk soil (30, 35, 36). Root-released organics are more likely to be decomposed by soil bacteria than carbon present in the bulk soil due to a lower degree of recalcitrance. Most of the root-released carbon is liberated near the root tip during initial root development (36). Therefore, the rhizosphere can be characterized as a region in soil with a high availability of carbon (albeit transient), as well as other nutrients such as N, P and S compounds. Structural cellular material liberated upon root death provides a more recalcitrant carbon source. Rhizosphere bacteria often show increased growth and activity due to the enhanced availability of organic carbon in the rhizosphere (28). Moreover, water flow in soil induced by roots may enhance bacterial movement towards them. Both mechanisms may promote cell-cell, cell-bacteriophage and cell-DNA contacts, which pictures the rhizosphere as an area in soil potentially conducive to genetic processes between soil bacteria.

12.3 Gene Transfer in Soil and Rhizosphere

All known gene transfer mechanisms, transformation, transduction and conjugation, are controlled by specific conditions that affect key bacterial parameters such as cellular physiology, population density and possibly cellular movement and establishment. These parameters affect the fate and survival of parent strains and potential excipients (transformants, transductants and transconjugants). Furthermore, specific effects of soil on transforming DNA and bacteriophages may play a significant role in determining transformation and transduction rates. There might, however, be a difference as to the extent these gene transfer mechanisms are affected by soil conditions, since they may have different requirements.

Since donor and excipient survival and activity is crucial for the impact gene transfer may have on soil populations, data on how bacterial survival is affected by soil factors are useful in the interpretation of results of gene transfer processes. The effects of factors such as soil type (texture), pH, moisture, temperature, clay and/or organic matter content and predatory protozoa have been reviewed (37, 38) and will not be treated here. Rather, gene transfer processes will be discussed with respect to the physical barriers to cell-cell, cell-phage and cell-DNA contacts, and nutritional limitations of soil. The nutritional upshift and putative alleviation of transport/contact barriers that bacteria may experience in the rhizosphere will also be discussed.

12.3.1 Transformation

For transformation to take place *in vitro*, bacterial cells have to develop competence, and transforming DNA has to contact the cells. For optimal transformation rates, specific conditions as to the nature of the transforming DNA and cellular competence development have to be met. These conditions are different for the different bacteria which are naturally transformable, such as *Bacillus, Streptococcus, Haemophilus, Pseudomonas and Acinetobacter* spp., and have been discussed elsewhere (39, 40, 41). Of potential ecological importance is the fact that many bacteria, e.g. bacilli, become competent when adverse conditions are prevalent. Also, at certain points of their life cycle bacteria make their DNA available for transformation via excretion into the extracellular environment (40). The soil solid phase might positively affect transformation since a multitude of adsorbable surfaces is available for cells and DNA that are in contact. However, soil might also negatively affect transformation of cells which are not in close contact with transforming DNA, since barriers to contact are also provided, e.g by the high surface areas of clays. It is at present unknown to what extent the presence of plant roots affects transformation rates in soil. However, it is conceivable that an alleviation of contact barriers occurs in soil sites temporarily influenced by roots, since the highly dynamic rhizosphere may provide conducive transformation conditions.

Studies on transformation in soil have mostly used disturbed soil microcosms, with or without prior sterilization of the soil. Simplified systems which cover some aspects of soil, such as sand columns have also been used (42). Data obtained in these systems have suggested that conditions conducive for transformation may occur in soil.

First, putative protection and persistence of DNA adsorbed to mineral surfaces and its availability for subsequent transformation was studied. In early studies in soil, DNA was suggested to be rapidly degraded in a (sandy) soil, however to persist in soil in association with montmorillonite clay (43). In a more recent *in vitro* study, DNA adsorbed to pure sand was shown to be resistant to DNase as opposed to its degradation in the unadsorded state (44). Furthermore, DNA bound to pure montmorillonite clay was also protected from degradation by DNase (45). Very recently, pUC18-ISP plasmid DNA added to soil was shown to persist for several weeks in 3 soils of different texture (46). Soil type affected both the rate of degradation and the availability of DNA for transformation of *Escherichia coli* after extraction from soil. The soil of finest texture protected the DNA better than the sandy soil. Together, these results corroborate the concept that by adsorbing to mineral surfaces, DNA is protected and persists in soil. The degree of protection is dependent on soil type, possibly due to varying clay, silt or sand surfaces available for adsorption and binding. Protection of DNA of lysing cells in soil via adsorption to mineral surfaces might be a mechanism by which genetic information becomes available for indigenous, naturally transformable, bacteria.

DNA adsorbed to clay or sand surfaces was shown to be available for transformation (42, 45, 46). Thus, *Bacillus subtilis* was transformable by DNA adsorbed to montmorillonite clay or mineral sand (42, 45). Also, pUC18::ISP DNA extracted from soil remained functional for transformation of *E. coli* for as long as 10 days in soil (46).

Moreover, bacterial cells, e.g. *Bacillus subtilis*, adsorbed to sand particles were shown to enter a state of enhanced competence; the frequency of transformation with chromosomal DNA and a *trp*C2 marker, of attached *B. subtilis* cells was 25- to 50-fold enhanced compared to that of unattached cells (42, 44). This would suggest that the presence of colonizable surfaces in soil stimulates competence development and transformation. It is however unknown whether this competence occurs for a wide range of bacteria.

Transformation of *Pseudomonas stutzeri* was shown to occur in soil extract with added nutrients, in particular carbon sources (47). Growth-limiting concentrations of single nutrients, including C, N and P sources, enhanced transformation. The heterogeneous distribution in soil of the different compounds needed for transformation and their presence at soil surfaces, however, was not dealt with.

Graham and Istock (48) first reported on cell contact transformation between bacilli in sterilized soil. Two introduced strains of *Bacillus subtilis* with different sets of chromosomal markers co-evolved into various types, with exchange of certain linkage blocks. Transformation was postulated as the transfer mechanism involved, even though the process was insensitive to DNase added to soil. Incubation in soil of one parent strain with DNA of the other strain also resulted in the appearance of transformants. Lee and Stotzky (49) also reported that bacilli co-introduced into soil exchanged chromosomal markers. Although these results suggested that cell contact dependent transformation mediated the transfers between added bacilli, transduction and conjugation cannot be completely ruled out.

Transformation in soil, however, may often escape detection due to its occurrence in soil microsites where conditions are suitable for cell-DNA contact. Overall transfer rates may be too low, against a background of bacteria in soil with the phenotype of the transferred trait. An early report on soil transformation (48) did not suffer from a background since sterile soil was used. Moreover, transfer over a whole soil system was studied instead of at the level of an individual soil microsite. Finally, the possible effect of plant roots in alleviating the barriers posed to cell-cell and cell-DNA contacts, and in enhancing bacterial activities, has been neglected so far. For these reasons, it is still difficult to provide an assessment of bacterial transformation in soil with regards to the evolution of soil populations and risks of spread of introduced recombinant DNA.

12.3.2 Transduction

In vitro transduction requires the presence of donor bacteria on which bacteriophages are propagated, of bacteriophage, and of recipient bacteria which are infected by phages carrying genes from the donor. Phage infection is often more prevalent in bacteria that are actively growing and metabolizing than in resting or stationary-phase cells. Adsorption of phages to bacterial cells generally requires the presence of bivalent cations like Mg^{2+}. At optimal conditions of cellular physiology and of the (abiotic) environment, the frequency of transduction is largely governed by the chance of contact between bacterial cells and phage particles, which is a function of cell and phage concentrations.

Transduction rates in soil are likely to be affected by the compartmentalization and physical separation of bacterial cells and bacteriophages in the soil pores, and by possible lack of ions promoting phage adsorption to cells. In addition, the relative immobility of both cells and phages due to the large surfaces present, blocks mixing and therefore cell-phage contacts. Compartmentalization could, however, also promote cell-phage contact in microsites where both players are present in densities sufficient for significant cell-phage contact rates. Further, the frequent lack of bacterial activity in soil may often preclude infection and, thus, transduction.

The chances for transduction in soil are enhanced by an abundance of phages in soil following a burst of infected host cells, which act as reservoirs of phages (50, 51). Artificial nutritional conditions promoting bacterial activity were conducive to phage propagation (51). Thus, indigenous bacteriophages of different *Bacillus* spp. significantly increased their titers in soil amended with nutrients but not in unamended soil (51). Furthermore, addition of such phages to two different soils which had been pre-sterilized and to which their host had been added (in the presence of nutrients) resulted in enhanced phage titers, of up to 10^{10} per g of dry soil, after a short incubation time in both soils (51). Herron and Wellington (52) also found initial propagation of the temperate actinophage KC301 on its host *Streptomyces lividans* when added to both sterile chitin- or starch-amended and non-sterile soils. The *Pseudomonas fluorescens* bacteriophage, ΦR2f. (53), added to soil seeded with the host strain, reduced the host population size in soil only in the presence of added nutrients (Smit et al., in preparation), suggesting that these phages reached sites in soil where host cells were present, but that host cell metabolism had to be stimulated for lysis to occur. Also, Stephens et al. (54) showed that a naturally occurring phage was responsible for the decline of introduced fluorescent pseudomonads in the rhizosphere of sugarbeet. This suggested that propagation of the indigenous phage at the expense of the introduced population took place in the relatively nutrient-rich rhizosphere. Propagation of bacteriophage on hosts present in soil, however, may be limited to host densities above a certain level, estimated to be around 10^4 cfu per ml or g (55).

Besides phage propagation and abundance in soil, the persistence of phages also influences the chances for transduction to occur. Phage persistence may vary depending on phage type and soil conditions. Thus, phages of different bacilli added to sterilized soil showed variable persistence at different temperatures (51). Phage survival was generally highest at the lowest temperature, 4 °C, at which phage titers hardly changed, whereas it was considerably lower at 28 °C and 37 °C. Soil type also affected phage survival; phage persisted poorly in a soil of low pH, but considerably better in a more neutral soil. Herron and Wellington (52) also showed that the actinophage KC301 steadily declined in soil following sporulation of the host. However, Germida (56) and Stotzky (57) showed that different bacteriophages and viruses may persist in soil and survive different physical and chemical stresses for prolonged periods of time even in the absence of hosts. Persistence of bacteriophages T2 and MS2 was affected by soil type (58); whereas, in some soils phage remained stable for 30 days, in others rapid inactivation in less than 5 days occurred. Soil temperature and texture also affected the survival of phages MS2 and PRD-1, in two soils (59). At increasing temperature, the inactivation rate was higher. Furthermore, the phage was better protected from inactivation in a clay loam than in sandy soils. Soil drying resulted in substantial phage inactivation. These data suggested that soil warming and drying adversely affect bacteriophage viability, and that soil texture affects phage persistence, enhancing survival in soils of finer texture.

The potential for the occurrence of transduction in soil has been demonstrated in soil microcosms (60, 61). *Escherichia coli* cells added to soil were shown to be reached by phage P1 carrying selectable markers, resulting in stable introduction of marker genes. Using a transducing lysate obtained from donor cells marked with auxotrophic markers and transposon Tn10, the frequency of transduction of *E. coli* in soil was 10^{-6} (60). A lysate consisting of phage particles carrying the selectable marker was apparently used by Zeph et al. (61), suggesting the process studied was more akin to phage conversion. Moreover, transduction of thiostrepton resistance by bacteriophage KC301 between streptomycetes in sterile starch- and chitin-amended soil was reported by Herron (62).

Transduction may thus occur in soil, in particular in microsites where conditions are conducive. The influence of the rhizosphere is still enigmatic, but nutrient stimulation of potential phage hosts may promote productive infections and thus putative transductional transfers. The frequencies of such transfers measured over a total soil system may, however, be extremely low. Therefore, it is difficult to assess to what extent transduction occurs in natural soil. There is as yet no conclusive evidence as to the impact transductional gene transfer has on soil bacteria with the phage concentrations found in natural soil.

12.3.3 Conjugation

Conjugal gene transfer *in vitro* depends on contact between donor and recipient cells under favourable abiotic conditions like temperature, pH, and nutrient availability. Often, cells adsorbed onto a surface efficiently form mating aggregates in the presence of dissolved nutrients. Transfer in soil is likely to be affected by the same factors, which in turn are governed by soil characteristics. Moreover, the availability of water in soil pores affects conjugal gene transfer, since a minimal water film surrounding cells in soil pores is probably required for most bacterial cells to remain physiologically active.

A large amount of data on conjugation between bacteria in soil has accumulated recently (e. g. 53, 63, 64, 65, 66, 67, 77). Data were obtained by either of two approaches. First, tracking similar plasmids in different rhizobia in soil (65) has provided evidence of presumably conjugal transfer. However, it is unclear whether the transfer event occurred in nodules or in the soil. Second, experiments with donor bacteria carrying transferable selectable markers, either accompanied or not by recipients, added to soil, have provided direct evidence of conjugal gene transfer (37, 53, 63, 66, 67, 68, 69, 70, 71, 77). Both plasmid-encoded (53, 66, 69) and chromosomal (63) genes have been shown to be transferable in natural soil. Transfers between introduced Gram-positive donor and recipient cells, e. g. different *Bacillus* spp. (69) and between introduced Gram-negative cells, e. g. between differentially-marked *Pseudomonas fluorescens* strains (66, 67) or from *Escherichia coli* to *Rhizobium fredii* (77) have been detected in soil. This topic has been reviewed (37). Here, we will focus on recent developments in conjugal gene transfer as affected by soil and rhizosphere.

Transfer of a derivative of the self-transmissible broad-host-range plasmid RP4 from *Pseudomonas fluorescens* to different Gram-negative members of the indigenous soil bacterial community was demonstrated in soil (53), highlighting the potential of soil bacteria to act as *in situ* recipients. Transfer in a sandy soil was most prominent under direct influence of wheat roots. Transfer was later shown to occur in 4 different soils, and the stimulative effect of the rhizosphere was dependent on soil type (72). In addition, mobilization of derivatives of plasmid RSF1010 from either *P. fluorescens* or *Escherichia coli* to Gram-negative indigenous bacteria was demonstrated in soil (63, 73). However, in spite of the finding that RSF1010 is transferable from Gram-negative to Gram-positive species (74), no such transfer was as yet observed in soil. Further, the conjugative transposon Tn916 which is transferable *in vitro* from Gram-positive to Gram-negative bacteria (75) was recently found to be transferred from introduced *Bacillus subtilis* to indigenous *Streptomyces* spp. in nutrient-amended soil (76). However, heterogramic transfer of Tn916 has not been observed in soil, even though it is able to transfer and express its selectable marker (resistance to tetracycline), in both Gram-positives and Gram-negatives.

Conjugal transfer in soil has been shown to be affected by many soil factors (37, 38, 67). Transconjugants could be detected in soil after the addition of nutrients (66, 68, 71), after soil sterilization which removes competitors for substrate (77) and in the presence of plant roots (66). In addition, the presence in soil of montmorillonite (64) or bentonite clay minerals (67) greatly enhanced conjugal transfer between introduced bacteria. Furthermore, conjugation rates were highest at moderate pH, whereas acid conditions did not permit transfer (37, 77). Soil organic matter content stimulated conjugal transfer in a sterile soil (77), but it lowered conjugation rates in the wheat rhizosphere under nonsterile conditions (66).

The presence of the soil solid phase is a dominating factor governing soil conjugal transfer, on the one hand by providing surfaces where bacterial mating aggregates can be formed and on the other hand by providing barriers to contacts between cells that are spatially separated. The drastic enhancement of conjugal transfer in the presence of bentonite clay (67, 69) may have been caused by the enhanced presence of sites for bacterial adsorption. The observation that transfer of the self-transmissible plasmid RP4p to indigenous bacteria was significantly enhanced in a clay soil as compared to a loamy sand soil supports this view (72). Alternatively, the enhancement of conjugal transfer in soil by montmorillonite might have been caused by the more neutral soil pH established by the clay (64, 78). That soil poses a barrier to cell-cell contact was shown recently (70); bacteria initially introduced into different soil portions which were subsequently mixed, were less able to transfer plasmid RP4 than cells added to the same soil portion, at different points in time. The presence of wheat roots alleviated the barrier effect of soil, allowing the detection of transconjugants also in the mixed soil portions.

The oligotrophic conditions in soil pose a second barrier to conjugation, since cellular energy is needed for a successful conjugal transfer event. Most bacteria in soil are in an almost permanent state of starvation and therefore probably unable to serve as donors or recipients in conjugal matings. Again, the influx of substrate into soil, either in the form of decaying plant material, from soil animals or in root exudates, can overcome this barrier, resulting in detectable conjugal gene transfer to indigenous bacteria in the rhizosphere (53, 72).

Recently, evidence was found for the occurence of conjugative plasmids in rhizosphere bacteria isolated from sugar beet roots that could mobilize Inc Q plasmids (79). The presence of such genetic elements in soil bacteria should be taken into account when genetically engineered bacteria are introduced into soil, since they might be responsible for enhancing the rate of transfer of heterologous DNA to indigenous bacteria.

12.4 Concluding Remarks

Current insight into bacterial genetics in soil suggests that the three known transfer mechanisms are in all likelihood operational in soil and rhizosphere. The most conclusive evidence is available for conjugation; effects of soil factors on conjugation have been extensively described (37). The occurrence and impact of transformation occurring in soil, primarily as a gene transfer mechanism between soil bacilli, is clearer now in the light of recent circumstantial evidence on the persistence of DNA in soil and the increased transformability of bacilli when adsorbed on mineral surfaces. Few studies report on the putative occurrence of transduction in soil, with one report describing phage conversion rather than transduction. Persistence and propagation of bacteriophages specific for soil bacteria in soil has been shown, leading to speculations on the role of phages in soil transductions.

Due to the large surface area of the solid matter and the gross oligotrophy, soil poses particular barriers to bacterial gene transfer. These barriers to gene transfer are (i) physical, i. e. the presence of structured particulate matter in soil impairs free mixing and movement of microorganisms, bacteriophages and DNA present, and (ii) nutritional, i. e. the transfer of genes is hampered due to the lack of energy or carbon sources needed for expression of cellular gene systems involved in transfer. Soil factors which reduce bacterial population densities, such as predation by protozoa, severe drought or freezing/thawing, may also affect gene transfer rates negatively by reducing the sizes of donor, recipient and excipient populations.

On the other hand, due to the presence of large surfaces, soil may potentially enhance gene transfer rates since gene transfers, particularly those via transformation and conjugation, are often stimulated at surfaces. The barriers to gene transfer may be alleviated in soil under conditions in which bacterial growth and activity is possible. A paradigm of such hot spots for bacterial activity is the rhizosphere. The enhanced nutrient input and water fluxes in the rhizosphere have been suggested to stimulate conjugal gene transfer between pseudomonads (66). Although the effect of plant roots on transformation and transduction in the rhizosphere is currently unknown, it is possible that both processes are stimulated due to effects similar to those promoting conjugation.

For an introduced GEM with chromosomally inserted heterologous DNA to successfully transfer this DNA to indigenous microbes by any one of the three transfer mechanisms, a sequence of events as outlined in Table 1 has to occur. Each event may be estimated to occur at a range of frequencies. A combination of such frequencies leads to the tentative conclusion that all frequencies are low to negligible (Table 1), and that a GEM of this type can be "gene transfer safely" released.

Table 1. Hypothetical Frequency of Transfer Event of Heterologous DNA Inserted in the Chromosome of a Bacterium Introduced into Soil[1].

Conjugation*	freq.*	Transduction	freq.	Transformation	freq.
GEMs and indigenous bacteria collide	10^{0}–10^{-6}	GEM and indigenous phage collide	10^{-3}–10^{-6}		10^{-1}–10^{-4}
GEM acquires plasmid	10^{-3}–10^{-9}	phage infects GEM	10^{-1}–10^{-8}	GEM excretes DNA	10^{-1}–10^{-4}
plasmid recruits heterologous gene	10^{-5}–10^{-8}	phage acquires heterologous gene	10^{-6}–10^{-8}	DNA persist	
GEM and indigenous recipient collide	10^{0}–10^{-6}	phage and indigenous recipient collide	10^{-3}–10^{-6}	DNA and indigenous bacterium collide	10^{-2}–10^{-4}
plasmid is transferred	10^{-1}–10^{-9}	phage infects	10^{-1}–10^{-7}	DNA taken up & stabilized in new host	10^{-4}–10^{-8}
gene is expressed	10^{0}–10^{-2}	gene is expressed	10^{0}–10^{-2}	gene is expressed	10^{0}–10^{-2}
total frequency	10^{-9}–10^{-40}		10^{-14}–10^{-37}		10^{-9}–10^{-22}

* freq.: estimated frequency of occurrence
[1] after Lacey and Stromberg (80)

One of the major difficulties in studying gene transfer events in soil is the rarity of the events studied as is depicted in Table 1. Transfers still detectable *in vitro* due to the lack of background, often escape detection in soil, due to the lower frequencies of transfer and the higher background found when applying selection. An additional complication is the lack of culturability of a major part of the bacterial population in soil, which impairs the detection of potential non-culturable excipients by conventional means.

An improved picture of gene transfer processes in soil is obtained in experiments aiming to track both the short-term fate of genetic material in soil, and the effects of selection on its long-term occurrence. The low-frequency transfer events which escape short-term detection due to the background in soil, may become detectable following selection for the transferred trait. In particular, transfer via transformation or transduction of selectable traits could thus become detectable. In addition, the conjugal transfers, eg. of conjugative transposon Tn916 or of broad-host-range mobilizable plasmid RSF1010, might have eluded detection and become detectable.

Acknowledgements

We thank Drs. L.S. van Overbeek and Dr. J.T. Trevors for critically reading this manuscript.

12.5 References

1. De Freitas, J.R. and Germida, J.J. Growth promotion of winter wheat by fluorescent pseudomonads under field conditions. Soil. Biol. Biochem. *24* (1992) 1137–1146.
2. Gaskins, M.H., Albrecht, S.L., Hubbel, D.H. Rhizosphere bacteria and their use to increase plant productivity: A review. Agric. Ecosyst. Environ. *12* (1985) 99–116.
3. Lugtenberg, B.J.J., and de Weger, L.A. Plant root colonization by *Pseudomonas* species. In: *Pseudomonas* Molecular biology and biotechnology, E. Galli, S. Silver, B. Witholt (Eds.), Am. Soc. Microbiol. Washington DC, 1992, pp. 13–19.
4. Keel, C., Wirthner, P., Oberhansli, T., Voisard, C., Burger, U., Haas, D., and Defago, G. Pseudomonads as antagonists of plant pathogens in the rhizosphere: role of the antibiotic 2,4-diacetylphloroglucinol in the suppression of black root rot of tobacco. Symbiosis *9* (1990) 327–341.
5. Lugtenberg, B.J.J., de Weger, L.A., Bennet, J.W. Microbial stimulation of plant growth and protection from disease. Curr. Opinion Biotechnol. *2* (1991) 457–464.
6. Schroth, M.N., Loper, J.E. and Hildebrand, D.C. Bacteria as biocontrol agents of plant disease. In: Current perspectives in microbiol ecology. M.J. Klug, C.A. Reddy, (Eds.). Am. Soc. Microbiol. Washington DC, 1984, pp. 362–369.
7. Alexander, M. Biodegradation of chemicals of environmental concern. Science *211* (1981) 132–138.
8. Brunner, W., Sutherland, F.H., Focht, D.D. Enhanced biodegradation of polychlorinated biphenyls in soil by analog enrichment and bacterial inoculation. J. Environ. Qual. *14* (1985) 324–328.
9. Pipke, R., Wagner-Döbler, I., Timmis, K.N. and Dwyer, D.F. Survival and function of a genetically engineered pseudomonad in aquatic sediment microcosms. Appl. Environ. Microbiol. *58* (1992) 1259–1265.
10. Doyle, J.D., Short, K.A., Stozky, G., King, R.J., Seidler, R.J. and Olsen, R.H. Ecologically significant effects of *Pseudomonas putida* PPO301 (pRO103) genetically engineered to degrade 2,4-dichlorophenoxyacetate on microbial populations and processes in soil. Can. J. Microbiol. *37* (1991) 682–691.

11. Duque, E., Ramos-Gonzales, M.I., Delgado, A., Contreras, A., Molin, S. and Ramos, J.L. Genetically engineered *Pseudomonas* strains for mineralization of aromatics: Survival, performance, gene transfer and biological containment. In: *Pseudomonas* Molecular biology and Biotechnology. E. Galli, S. Silver, B. Witholt, (Eds.). Am. Soc. Microbiol. Washington DC, 1992, pp. 429–437.
12. Fenton, A.M., Stephens, P.M., Crowley, J., O'Callaghan, M. and O'Gara, F. Exploitation of gene(s) involved in 2,4-diacetylphloroglucinol biosynthesis to confer a new biocontrol capability to a *Pseudomonas* strain. Appl. Environ. Microbiol. *58* (1992) 3873–3878.
13. Lindow, S.E. Ecology of *Pseudomonas syringae* relevant to the field use of Ice$^-$ deletion mutants constructed *in vitro* for plant frost control. In: Engineered organisms in the environment. H.O. Halvorson, D. Pramer, M. Rogul, (Eds.). Am. Soc. Microbiol. Washington DC, 1985, pp. 23–25.
14. Ramos, J.L., Duque, E., Ramos-Gonzales, M.I. Survival in soils of a herbicide-resistant *Pseudomonas putida* strain bearing a recombinant TOL plasmid. Appl. Environ. Microbiol. *57* (1991) 260–266.
15. Skøt, L., Harrison, S.P., Nath, A., Mytton, L.R. and Clifford, B.C. Expression of insecticidal activity in *Rhizobium* containing the δ-endotoxin gene cloned from *Bacillus thuringiensis* subsp. *tenebrionis.* Plant and Soil *127* (1990) 285–295.
16. Van Elsas, J.D., van Overbeek, L.S., Feldmann, A.M., Dullemans, A.M. and de Leeuw, O. Survival of a genetically engineered *Pseudomonas fluorescens* strain in soil in competition with the parent strain. FEMS Microbiol. Ecol. *85* (1991) 53–64.
17. Waalwijk, C., Dullemans, A. and Maat, C. Construction of a bioinsecticidal rhizosphere isolate of *Pseudomonas fluorescens.* FEMS Microbiol. Lett. *77* (1991) 257–264.
18. Griffiths, R.F. Dealing with risk. Manchester University Press, Manchester, 1981.
19. Levin, M.A. and Strauss, H.S. Risk assessment in genetic engineering. McGraw-Hill, New York, 1990.
20. Reanney, D.C. Gene transfer as a mechanism of microbial evolution. BioScience *27* (1977) 340–344.
21. Wellington, E.M.H. and van Elsas, J.D. Gene Transfer Between Microorganisms in the Natural Environment. Pergamon Press, London, 1992.
22. Fry, J.C. and Day, M.J. Bacterial Genetics in the Natural Environment. Chapman and Hall, London, 1990.
23. Levy, S.B. and Miller, R. Gene Transfer in the Environment. McGraw-Hill, New York, 1989.
24. Smiles, D.E. Aspects of the physical environment of soil organisms. Biol. Fertil. Soils *6* (1988) 204–215.
25. Van Veen, J.A. and Van Elsas, J.D. Impact of soil structure and texture on activity and dynamics of the soil microbial population. In: Perspectives in Microbial Ecology. Megusar, F. and Gantar, M., (Eds.). Slovena Society for Microbiology, Ljubljana, 1986, pp. 481–488.
26. Hattori, T. and Hattori, R. The physical environment in soil microbiology: an attempt to extend principles of microbiology to soil microorganisms. CRC Critical Reviews in Microbiology *4* (1976) 423–461.
27. Trevors, J.T., van Elsas, J.D., van Overbeek, L.S. and Starodub, M.E. Transport of a genetically engineered *Pseudomonas fluorescens* strain through a soil microcosm. Appl. Environ. Microbiol. *56* (1990) 401–408.
28. Foster, R.C. Microenvironments of soil microorganisms. Biol. Fertil. Soils *6* (1988) 189–203.
29. Williams, S.T. Oligotrophy in soil: fact or fiction? In: Bacteria in the natural environment: the effect of nutrient conditions. Fletcher, M. and Floodgate, G., (Eds.). Acad. Press, London, 1985, pp. 81–110.
30. Van Elsas, J.D. and van Overbeek, L.S. Bacterial responses to soil stimuli. In: Starvation of bacteria. Kjelleberg, S. (Ed.) Plenum Press, New York, 1993, 55–79.
31. Gray, T.R.G. and Williams, S.T. Microbial productivity in soil. Symp. Soc. Gen. Microbiol. *21* (1971) 255–286.
32. Shields, J.A., Paul, E.A., Lowe, W.E., Parkinson, D. Turnover of microbial tissue in soil under field conditions. Soil Biol. Biochem. *5* (1973) 753–764.
33. Curl, A.E. and Truelove, B. The rhizosphere. Springer Verlag, Berlin, 1986.
34. Lynch, J.M. and Whipps, J.M. Substrate flow in the rhizosphere. Plant and Soil *129* (1990) 1–10.
35. Newman, E.I. The rhizosphere: carbon sources and microbial populations. In: Ecological interactions in soil. Fitter, A.H., (Ed.). Spec. Publ. No. 4 of the British Ecological Society, Blackwell, Oxford, 1985, pp. 107–121.
36. Trofymow, J.A., Coleman, D.C. and Cambardella, C. Rates of rhizodeposition and ammonium depletion in the rhizosphere of axenic oat roots. Plant and Soil *97* (1987) 333–344.

37. Stotzky, G. Gene transfer among bacteria in soil. In: Gene Transfer in the Natural Environment. Levy, S.B. and Miller, R.V. (Eds.). McGraw-Hill, N.Y., 1989, pp. 165-222.
38. Van Elsas, J.D. and Trevors, J.T. Plasmid transfer to indigenous bacteria in soil and rhizosphere: problems and perspectives. In: Bacterial genetics in Natural Environments. Fry, J.C. and Day, M.J. (Eds.). Chapman and Hall, London, 1990, pp. 188-199.
39. Stewart, G. The Mechanism of Natural Transformation. In: Gene Transfer in the Natural Environment. Levy, S.B. and Miller, R. (Eds.). McGraw-Hill, N.Y., 1989, pp. 139-164.
40. Stewart, G. Transformation in Natural Environments. In: Genetic interactions between bacteria in the environment. Wellington, E.M.H. and van Elsas, J.D. (Eds.). Pergamon Press, London, 1992, pp. 216-234.
41. Stewart, G.J. and Sinigalliano, C.D. Detection of horizontal gene transfer by natural transformation in native and introduced species of bacteria in marine and synthetic sediments. Appl. Environ. Microbiol. *56* (1990) 1818-1824.
42. Lorenz, M.G., Aardema, B.W., Wackernagel, W. Highly efficient genetic transformation of *Bacillus subtilis* attached to sand grains. J. Gen. Microbiol. *134* (1988) 107-112.
43. Greaves, M.P. and Wilson, M.J. The degradation of nucleic acids and montmorillonite-nucleic-acid complexes by soil microorganisms. Soil Biol. Biochem. *2* (1970) 257-268.
44. Lorenz, M.G. and Wackernagel, W. Adsorption of DNA to sand and variable degradation rates of adsorbed DNA. Appl. Environ. Microbiol. *53* (1987) 2948-2952.
45. Khanna, M. and Stotzky, G. Transformation of *Bacillus subtilis* by DNA bound on montmorillonite and effect of DNase on the transforming ability of bound DNA. Appl. Environ. Microbiol. *58* (1992) 1930-1939.
46. Romanowski, G., Lorenz, M.G., Sayler, G.S. and Wackernagel, W. Persistence of free plasmid DNA in soil monitored by various methods, including a transformation assay. Appl. Environ. Microbiol. *58* (1992) 3012-3019.
47. Lorenz, M.G. and Wackernagel, W. High frequency of natural genetic transformation of *Pseudomonas stutzeri* in soil extract supplemented with a carbon/energy and phosphorus source. Appl. Environ. Microbiol. *57* (1991) 1246-1251.
48. Graham, J.B. and Istock, C.A. Genetic exchange in *Bacillus subtilis* in soil. Mol. Gen. Genet. *116* (1978) 287-298.
49. Lee, G.-H. and Stotzky, G. Transformation is a mechanism of gene transfer in soil. Kor. J. Microbiol. *28* (1990) 210-218.
50. Reanney, D.C., Gowland, P.C. and Slater, J.H. Genetic interactions among microbial communities. Symp. Soc. Gen. Microbiol. *34* (1983) 379-421.
51. Van Elsas, J.D. Ecologia e taxonomia de *Bacillus* e seus bacteriófagos. Ph.D. thesis, 1983, UFRJ, Brasil.
52. Herron, P. and Wellington, E.M.H. New method for extraction of streptomycete spores from soil and application to the study of lysogeny in sterile amended and nonsterile soil. Appl. Environ. Microbiol. *56* (1990) 1406-1412.
53. Smit, E., Van Elsas, J.D., van Veen J.A. and de Vos, W.M. Detection of plasmid transfer from *Pseudomonas fluorescens* to indigenous bacteria in soil by using phage FR2f. for donor counterselection. Appl. Environ. Microbiol. *57* (1991) 3482-3488.
54. Stephens, P.M., O'Sullivan, M. and O'Gara, F. Effect of bacteriophage on colonization of sugar beet roots by fluorescent *Pseudomonas* spp. Appl. Environ. Microbiol. *53* (1987) 1164-1167.
55. Wiggins, B.A. and Alexander, M. Minimum bacterial density for bacteriophage replication: implications for significance of bacteriophages in natural ecosystems. Appl. Environ. Microbiol. *49* (1985) 19-23.
56. Germida, J.J. Population dynamics of *Azospirillum brasilense* and its bacteriophage in soil. Plant and Soil *90* (1986) 117-128.
57. Stotzky, G. Surface interactions between clay minerals and microbes, viruses and soluble organics, and the probable importance of these interactions to the ecology of microbes. In: Microbial adhesion to surfaces, Berkeley, R.C.W., Lynch, J.M., Melling, J., Rutter, P.R. and Vincent, B. (Eds.). Ellis Horwood Ltd., Chichester, 1980, pp. 231-247.
58. Hurst, C.J., Gerba, C.P. and Cech, I. Effect of environmental variables and soil characteristics on virus survival in soil. Appl. Environ. Microbiol. *40* (1980) 1067-1079.
59. Straub, T.M., Pepper, I. L. and Gerba, C.P. Persistence of viruses in desert soils amended with anaerobically digested sewage sludge. Appl. Environ. Microbiol. *58* (1992) 636-641.

60. Germida, J.J. and Khachatourians, G.G. Transduction of *Escherichia coli* in soil. Can. J. Microbiol. *34* (1988) 190–193.
61. Zeph, L.R., Onaga, M.A. and Stotzky, G. Transduction of *Escherichia coli* by bacteriophage P1 in soil. Appl. Environ. Microbiol. *54* (1988) 1731–1737.
62. Herron, P. Interactions between actinophage and streptomycetes in soil. Ph.D. thesis, 1991, U. of Warwick, Coventry, UK.
63. Henschke, R.B. and Schmidt, F.R.J. Plasmid mobilization from genetically engineered bacteria to members of the indigenous soil microflora *in situ*. Curr. Microbiol. *20* (1990) 105–110.
64. Krasovsky, V.N. and Stotzky, G. Conjugation and genetic recombination in *Escherichia coli* in sterile and non-sterile soil. Soil Biol. Biochem *19* (1987) 631–638.
65. Schofield, P.R., Gibson, A.H., Dudman, W.F. and Watson, J.M. Evidence for genetic exchange and recombination of *Rhizobium* symbiotic plasmids in a soil population. Appl. Environ. Microbiol. *53* (1987) 2942–2947.
66. Van Elsas, J.D., Trevors, J.T. and Starodub, M.-E. Bacterial conjugation between pseudomonads in the rhizosphere of wheat. FEMS Microbiol. Ecol. *53* (1988a) 299–306.
67. Van Elsas, J.D., Trevors, J.T., and Starodub, M.E. Plasmid transfer in soil and rhizosphere. In: Risk Assessment for Deliberate Releases. Klingmuller, W. (Ed.). Springer Verlag, Heidelberg, 1988b, pp. 89–99.
68. Top, E., Mergeay, M., Springael, D. and Verstraete, W. Gene escape model: transfer of heavy metal resistance genes from *Escherichia coli* to *Alcaligenes eutrophus* on agar plates and in soil samples. Appl. Environ. Microbiol. *56* (1990) 2471–2479.
69. Van Elsas, J.D., Govaert, J.M. and Van Veen, J.A. Transfer of plasmid pFT30 between bacilli in soil as influenced by bacterial population dynamics and soil conditions. Soil Biol. Biochem. *19* (1987) 639–647.
70. Van Elsas, J.D. Trevors, J.T. Starodub, M.E. and van Overbeek, L.S. Transfer of plasmid RP4 between pseudomonads after introduction into soil; influence of spatial and temporal aspects of inoculation. FEMS Microbiol. Ecol. *73* (1990) 1–12.
71. Wellington, E.M.H., Cresswell, N. and Saunders, V.A. Growth and survival of Streptomycete inoculants and extent of plasmid transfer in sterile and nonsterile soil. Appl. Environ. Microbiol. *56* (1990) 1413–1419.
72. Richaume, A., Smit, E. Faurie, G. and van Elsas, J.D. Influence of soil type on the transfer of plasmid RP4p from *Pseudomonas fluorescens* to introduced recipients and to indigenous bacteria. FEMS Microbiol. Ecol. *101* (1992) 281–292.
73. Smit, E., Venne, D. and van Elsas, J.D. Mobilization of a recombinant Inc Q plasmid between bacteria on agar and in soil via CO-transfer or retro-transfer. Appl. Environ. Microbiol. *59* (1993) 2257–2263.
74. Gormley, E.P. and Davies, J. Transfer of plasmid RSF1010 by conjugation from *Escherichia coli* to *Streptomyces lividans* and *Mycobacterium smegmatis*. J. Bacteriol. *173* (1991) 6705–6708.
75. Bettram, J., Strätz, H. and Dürre, P. Natural transfer of conjugative transposon Tn916 between Gram-positive and Gram-negative bacteria. J. Bacteriol. *173* (1991) 443–448.
76. Natarajan, M.R. and Oriel, P. Transfer of transposon Tn916 from *Bacillus subtilis* into a natural soil population. Appl. Environ. Microbiol. *58* (1992) 2701–2703.
77. Richaume, A., Angle, J.S. and Sadowski, M.J. Influence of soil variables on in situ plasmid transfer from *Escherichia coli* to *Rhizobium fredii*. Appl. Environ. Microbiol. *55* (1989) 1730–1734.
78. Weinberg, S.R. and Stotzky, G. Conjugation and genetic recombination of *Escherichia coli* in soil. Soil Biol. Biochem. *4* (1972) 171–180.
79. Bailey, M.J., Kobayashi, N., Lilley, A.K., Powell, B.J. and Thompson, I.P. Assessment for the potential for gene transfer in the phytosphere of sugar beet. In: Gene transfers and Environment. Gauthier M.J. (Ed.). Springer Verlag Berlin. 1992 143–148.
80. Lacey, G.K. and Stromberg, V.K. Pre-release microcosm tests with a genetically engineered plant pathogen. In: Biological monitoring of genetically engineered plants and microbes. Proceedings of the Kiawah Island Conference, Mackenzie, D.R. and Henry, S.C., (Eds.). Agricultural Research Institute Bethesda, Bethesda, MD., 1991, pp. 81–98.

13 European Community Regulation for the Use and Release of Genetically Modified Organisms (GMOs) in the Environment

Marco P. Nuti, Andrea Squartini and Alessio Giacomini

13.1 Introduction

The agricultural use of plants and animals in open environments goes along with man's history. Pure bacterial cultures have been used for agricultural purposes, for almost a century (1). Inoculation of legumes with rhizobia represents the longest recorded usage of microbes in open land. However, during the last two decades other soil/seed microbial inoculants have been described including plant protectants (mainly biocontrol agents of plant pests and diseases) and plant promoting rhizobacteria and rhizofungi. The use of microbes for agricultural purposes extends over a considerable area (Table 1) and is expected to increase in the framework of land managerial practices such as sustainable agriculture and organic farming.

Tab. 1. Relevant Contribution of Microbial Inoculants to Food Production and Soil Management

Microorganism	Plant inoculated	Reference	Area (ha) inoculated (*)
Anabaena-Azolla	rice	(2)	$>2\times10^6$
Azospirillum brasilense, *A. lipoferum*	cereals (Mainly *Triticum durum*, *Zea mays*, grasses)	(3) (4)	$1.5{-}2\times10^5$
Bradyrhizobium japonicum, *Rhizobium* spp.	forage and grain legumes	(5) (6)	$1.2{-}2.4\times10^7$
Frankia spp.	non-leguminous trees (*Alnus, Casuarina, Ceanothus, Hippophae, Myrica*)	(7)	$0.5{-}1\times10^3$
mycorhizal fungi (*Boletus* spp., *Pisolithus tinctorius, Tuber aestivum, T. albidum, T. magnatum, T. melanosporum*	forest trees and bushes (*Abies, Carpinus, Cedrus, Cistus, Corylus, Ostrya, Pinus, Populus, Quercus, Salix, Tilia)*	(8) (9) (10)	$0.5{-}1\times10^5$
Bacillus subtilis, B. thuringensis, Phialophora sp., *Pseudomonas fluorescens, Trichoderma viride*	crop plants and trees	(11)	$0.5{-}1\times10^5$

(*) estimated average area inoculated on annual basis

Modern biotechnology, and in particular the advancements in gene technology, have provided sufficient background for the successful introduction of recombinant (rDNA) biopharmaceutical products in the 1980's, including human insulin and α-interferon, hepatitis B vaccine, tissue plasminogen activator and erythropoietin. Following these achievements, rDNA products and live genetically modified organisms (GMOs) (12) are being introduced for both animal health care and crop agriculture (13). Within the first group fall the hormones that make livestock grow faster and cows produce more milk, as well as vaccines like those against porcine parvovirus or transmissable gastroenteritis virus. Live GMOs for crop agriculture include crop plants made resistant to pests (insects, nematodes), diseases (fungal, viral) and safe herbicides (e.g. glyphosate-based) along with seed/plant microbial inoculants to be used as biofertilizers, biopesticides or phytostimulators. A selected list of plants and microorganisms, which were genetically modified for improved agricultural use, is reported in Tables 2 and 3 (for bacteria, see also ref. 13).

Gene technology is one example of a new technology that raised concerns in the scientific community and in the public perception of it. Apparently, scientific concerns arose mainly because of uncertainty about gene transfer between phylogenetically distant organisms, while public concerns arose from unduly publicized descriptions of rDNA techniques as being "totipotent" or intrinsically hazardous for

Tab. 2. Selected Examples of Plants, whose traits were improved through Genetic Modification

Plant	Altered Trait and Purpose	Reference
Arabidopsis thaliana	insertion of genes for an acyl-carrier protein (BTE) from *Umbellularia californica* - medium-chain fatty acid production	(14)
Corn	introduction of synthetic gene encoding truncated CrylA(b) protein from *B. thuringensis* – resistance to *Ostrinia nubilalis*	(15)
Cucumber	introduction of coat protein gene of a non-aphid transmissable strain of cucumber mosaic virus (CMV) – resistance to CMV	(16)
Potato	insertion of cyclodextrin glycosyltransferase gene from *Klebsiella* – production of cyclodextrins	(17)
	insertion of PVX coat protein gene – increased resistance to potato virus X	(18)
Rice	introduction of a gene for resistance to phosphinothricin (PPT) – resistance to commercial herbicides containing PPT (e.g. Basta®)	(19)
Sugarbeet	introduction of bialaphos resistance gene (bar) – resistance to glufosinate-ammonium herbicide	(20)
Tobacco	introduction of a gene for a phaseolotoxin-resistant ornithyl transcarbamylase – resistance to *Pseudomonas syringae* pv. *phaseolicola*	(21)
Tomato	inhibition of polygalacturonase gene expression by antisense RNA – ripening control	(22)
Wheat	introduction of bialaphos resistance gene (bar) – resistance to the commercial herbicide Basta®	(23)

Tab. 3. Selected Examples of Microbial Strains, for Agricultural Use, whose traits were improved through Genetic Modification

Microorganism	Altered Trait and Purpose	Reference
Agrobacterium	deletion of *tra* genes of pAgK84 – biological control of crown gall	(24)
Alcaligenes eutrophus	heavy metal resistance and chloro-aromatics degradation – enhanced biodegradation for polluted soil reclamation	(25) (26)
Baculovirus	polyhedrin gene and transcription promoter removal – self-destructive virus	(27)
Clavibacter xyli	delta endotoxin gene from *B. thuringensis* subsp. *kurstaki* – prevention of insect damage in cultivated crops	(28)
Bradyrhizobium japonicum	additional copies of *nif* – increased N_2 fixation	(29)
Pseudomonas syringae	deletion of *ina* (ice nucleation gene) – control of frost damage to plants	(30)
Rhizobium meliloti	additional copies of *nif* or dicarboxylate transport *dct* – increased N_2 fixation	(31)

humans and environment. The two latter concepts were profoundly misleading as gene technology poses risks which are essentially the same in nature as those posed by conventional biotechnology and plant breeding, risks that can also be assessed in a similar way (12, 32). However, the installation of a sound regulatory framework can offer a sensible approach to overcome both emotional and scientific concerns.

13.2 The International Regulatory Framework

Several countries have adopted precautionary safety measures since the mid 1970's, when guidelines for laboratory work were first published. The OECD Blue Book (33) later provided a number of principles for safe handling of GMOs in open environments, and OECD still represents an international forum to discuss the relevant issues pertaining to biosafety and field testing of GMOs.

In the USA, the Plant Quarantine Act (PQA), the Federal Plant Pest Act (FPPA), the National Environmental Policy Act (NEPA) provide the regulatory framework, and since June 1986 the Animal and Plant Health Inspection Service (APHIS) of the U.S. Department of Agriculture (USDA) has granted approvals for field tests and product licensing of a wide variety of genetically engineered organisms and products thereof. When appropriate, the Environment Protection Agency (EPA) becomes involved as an implementing agency. A user's guide for biotechnology permits gives detailed assistance in the U.S. for field testing GMOs (34). A recent federal oversight revision in the U.S. has led to a modified policy on planned introductions of biotechnology products into the environment (35). Other countries have this matter regulated by specific Acts (e.g. Canada with the Environment Protection Act) or guidelines (e.g. Japan, New Zealand, Australia), implemented by Ministries or Governmental Bodies (36).

The above regulations are based on a case-by-case evaluation using statutes/guidelines intended to protect plants, animal health, and the human environment. Being directed to the assessment of the characteristics of the product, and not of the method by which the GMO has acquired the new traits, these regulations can be defined as "product-oriented".

13.3 The European Community Regulation

The European Community has adopted stringent precautionary safety measures for the deliberate release into the environment of genetically modified organisms. The regulatory framework to be implemented in all Member States is provided by the Council Decision of 23 April 1990 (Directive 90/220/EC), published in the Official Journal of the European Communities no. L117, vol. 33 on the 8th May 1990, concerning the deliberate release into the environment of genetically modified microorganisms. The Council Decision of 4 November 1991 (91/596/EC) published in the Official Journal no. L322 vol. 34 on the 23rd November 1991 concerns the summary notification information format referred to in Article 9 of the previous Directive.

According to Directive 90/220/EC, a GMO is defined as "an organism in which the genetic material has been altered in a way that does not occur naturally by mating and/or natural recombination". Therefore, GMOs excluded from the Directive are those organisms which are obtained through mutagenesis, and plants obtained by gene fusion, if the same can be obtained by traditional methods. The main features of this EC regulation include the following: (i) an environmental risk assessment must precede any release, (ii) no release can be carried out without the consent of the Competent Authorities, (iii) for experimental releases, national approval procedures are needed, (iv) for commercial releases, Community approval procedures are needed; once approved, products can freely circulate throughout the Community. To provide the environmental risk assessment, the requested information includes (a) the characteristic of the donor and recipient organisms, vector and GMO, (b) the conditions of release and of the receiving environment, (c) the interaction with the environment, namely the survival, multiplication, dissemination and the environmental impact of the GMO, (d) the monitoring techniques, waste treatment and emergency responses.

Not all Member States of the European Community have completed the implementation of the Directive 90/220. At the end of March 1993, the various member states can be grouped as follows:

1. those having adopted all the measures they plan to take, and consider they have implemented the Directives 90/220 and 91/596 (Denmark, Germany, Netherlands, United Kingdom);
2. those having adopted the framework legislation but are in the process of finalising the detailed regulation for full implementation (France, Ireland, Italy, Portugal);

the process is expected to be completed between April and July 1993; Belgium could be added to this group, as legislation has been adopted in one region, and is about to be adopted in the two other regions;
3. those where draft laws are being examined by the decision-taking Body, expected to be finished by the end of 1993 (Spain, Greece and Luxembourg).

All the Member States have appointed the Competent Authority (C.A.) and have an administrative structure to handle notifications. The C.A. are represented by the Ministry of Environment (Denmark, Greece, Ireland, Netherlands, Portugal, Spain), by the Ministry of Environment jointly with Ministry of Agriculture (France), by Health and Safety Executive (UK), by the Ministry of Health (Germany, Italy, Luxembourg), or by the Inst. of Health and Environment (Belgium).

Overall, the EC regulation requires an assessment of risks posed by GMOs, taking into consideration the method through which the genetic modification has been achieved. Therefore, this regulation can be defined as "technology-oriented".

13.4 Biosafety Results of Field Tests of GMOs

It is important, particularly for those countries having adopted stringent precautionary safety legislation for the use and commercialisation of GMOs, to learn from the ca. six hundred field tests. It is also important for them to assess whether there are key issues to which the scientific community can provide further insight. In Tables 4-6 the biosafety results of field tests and genetically modified vaccines, viruses, bacteria and plants are summarized. In general, these results seem to indicate that the behaviour of the GMOs in the environment does not differ significantly from that of the corresponding wild-types, unless the modified genetic trait is specifically designed to produce a different impact: for instance, a virus-resistant tomato will behave differently from its parental organism in that it will be resistant to the viral attack, or the recombinant anti-rabies vaccine will behave differently from the conventional one in that the recombinant vaccinia virus is not transmitted to rodents whereas live rabies vaccine can be transmitted to rodents.

Perhaps microbes seem to pose more subtle questions, compared to plants and animals, in particular when an environmental impact analysis is requested. However, it should be noted that nowadays molecular microbial ecology offers detection and monitoring methods which are substantially improved with respect to conventional microbiological techniques (37).

Irrespective of the organism to be used in the environment, it is unfortunate that in the existing technology-oriented regulation a definition of what can be considered "environmentally safe" is lacking. However, in product-oriented regulations, partial deregulation (35, 38) allows field releases of GMOs to be undertaken when findings of no significant impact are made available by the appropriate implementing Agency (e.g. USDA-APHIS in U.S.A.; an example is reported in ref. 28). Surprisingly, in the EC Directive even the concept of environment is not defined. This could lead to uneven interpretation by national competent authorities or possibly conflict-

Tab. 4. Biosafety Results of Field Tests of GMOs* (Vaccines and Viruses)

Major Safety Issue	GMO released	Results
• Survival • Effect of modified traits	• Salmonella live vaccine (5,000 sheep over 80,000)	• None of the 75,000 untreated sheep became infected: no pathogens were shed
	• Vaccinia antirabies vaccine (area = 10,000 Km^2)	• Disease disappeared from most of the initially contaminated area
	• Baculovirus AcNPv (*Autographa californica* Nuclear Polyedrosis virus)	• Target species were affected and non-target species were unaffected • Scorpion toxin is effective and humans unlikely to encounter damaging doses of killed larvae or virus • Inactivation of residual viruses at the release site is possible

* Data taken from (12)

Tab. 5. Biosafety Results of Field Tests of GMOs* (Plants)

Major Safety Issue	GMO released	Results
• Gene stability • Gene transfer • Pollen/seed dissemination • Effects of modified trait	• Potato	• No evidence of hybridization with *S. nigrum* and *S. dulcamara* • Pollen remains within 20 m distance • No unexpected effects including any form of pleiotropy or insertional mutagenesis
	• Sugarbeet	• Traits introduced by conventional breeding are comparable to those introduced via genetic modification
	• Brassica	• Level of interspecific crosses is low and infertility of progeny is high • Carry-over effects are possible
	Others (tomato, tobacco, etc.) for virus resistance traits	• China (500 ha); foreign genes appear to be stable over 3-year period and different environments

* Data taken from (12)

Tab. 6. Biosafety Results of Field Tests of GMOs* (Bacteria)

Major safety issue	GMO released	Results
• Persistence • Dissemination • Population dynamics • Competition • Community effects	• *Agrobacterium*	• Biological control of gall disease is effective without transfer of resistance genes to pathogenic agrobacteria
	• *Rhizobium, Bradyrhizobium*	• Improved symbiotic performances • Lower detection limit is below 1 CFU × g of soil • Improved identification methods are available • Stable insertion and maintenance of modified traits is possible

* Data taken from (12, 13)

ing provisions. Therefore, it seems appropriate in the EC that one continues to assess the perceived levels of risks posed by GMOs by determining if there is any increase over acceptable "natural" levels. Once a significant increase is discovered, sensible regulations should allow one to balance the risk against the benefits obtainable from the use of the GMO (for instance, agriculture *per se* is an environment impacting/disrupting activity, but still is retained as a primordial tool for food production).

13.5 Concluding Remarks

There is a general consensus within the international scientific community that risks posed by rDNA technology are essentially the same in nature as those posed by conventional breeding and that rDNA-modified organisms are not inherently risky or unpredictable (12, 32). This view relies also on the evidence that no adverse consequences have resulted from work with rDNA techniques during the last twenty years in laboratory/contained environments, and from about six hundred field releases of GMOs. The very same conceptual basis allows the use of rDNA techniques on humans (there are 38 human patients being clinically treated by gene therapy as of March 1993; M. Scarpa, pers. comm. and ref. 39). The above position seems to be in contrast with the fundamentals of technology-oriented regulations, and in agreement with product-oriented regulations. It would be desirable to overcome the conceptual conflict between the two types of regulations. For example, according to the U.S. view towards European biotechnology there is "an hostile political attitude reflected in incoherent and adversarial regulatory framework". Also for a number of EC industries (40) the regulatory framework for placing a product on the market "must be based on the assessment of the characteristics of the final product and not on the specific technology used in the production process". If biotechnology has to play a role in the European Community for the development of areas such as healthy food production, bioremediation and environment protection, let us hope that sensible regulations be adopted, or existing regulations be modified, within the spirit and the letter of article 130F of the Single European Act.

Acknowledgements

This work has been carried out as part of the contracts EC-BRIDGE BIOT-CT91-0283 and EC-COMETT II (project BIOMERIT).

13.6 References

1. Nobbe, P. Hiltner, L. Die Boden-Impfung für Leguminosen mit Reincultivierten Bakterien. (1896) Druck von A.A. Wagner, Hoechst am Main.
2. Boussiba, S. Nitrogen fixing cyanobacteria: potential use. In: "Nitrogen Fixation", M. Polsinelli, R. Materassi, and M. Vincenzini (Eds.). Kluwer Acad. Publ., 1991, pp. 487-490.

3. Okon Y., Fallik, E., Sarig, S., Yahalom, E., Tal, S. Plant growth promoting effects of *Azospirillum*. In: "Nitrogen Fixation: Hundreds Years After" H. Bothe, F.J. de Bruijn, W.E. Newton (Eds.). Gustav fisher Publ, 1988, pp. 741-746.
4. Nuti, M.P., Rubboli, P. Preliminary trials of field release of *Azospirillum brasilense* as inoculant in northern Italy. In: "Risk assessment for deliberate release" W. Klingmuller (Ed.). Springer-Verlag, 1987, pp. 46-49.
5. Vincent, J.M. Nitrogen fixation in legumes. Academic Press, 1982, pp. 1-288.
6. Nuti, M.P. & Casella, S. Advances in the utilization of rhizobia in arid environments. Arid Soil Res. Rehabil. *3* (1989) 243-258.
7. Schwintzer, C.R., Tjepkema J.D. The biology of *Frankia* and actinorhizal plants. Academic Press, San Diego, 1990.
8. Bonfante-Fasolo, P., Perotto, S. La cooperazione tra piante e funghi simbionti. Le Scienze *284* (1992) 34-44.
9. Giovannetti, G. Patent I-1 128 367, 1986.
10. Mischiati, P., Fontana, A. *In vitro* culture of *Tuber magnatum* mycelium isolated from mycorrhizas. Mycol. Res. *97*(1): (1993) 40-44.
11. Lynch, J.M., Hobbie, J.E. (Eds.). Microorganisms in action: concepts and applications in microbial ecology, Blackwell Sci. Publ., 1988.
12. Casper, R., Landsmann, J. (Eds.). The biosafety results of field tests of genetically modified plants and microorganisms. Biol. Bund. Land Fortwirtsch., Braunschweig, Germany, 1992.
13. Lindow, S.E., Panopoulos, N.J. and McFarland, B.L. Genetic engineering of bacteria from managed and natural habitats. Science *244* (1989) 1300-1307.
14. Voelker, T.A., Worrell, A.C., Anderson, L., Bleibaum, J., Fan, C., Hawkins D.J., Radke, S.E., Maelor Davies, H. Fatty acid biosynthesis redirected to medium chains in transgenic oilseed plants. Science *257* (1992) 72-74.
15. Koziel, M.G. et al. Field performance of Elite transgenic maize plants expressing an insecticidal protein derived from *Bacillus thuringensis*. Bio/Technology *11* (1993) 194-200.
16. Gonsalves, D., Chee, P., Providenti, R., Seem, R., Slightom, J.L. Comparison of coat protein-mediated and genetically derived resistance in Cucumber to infections by CMV under field conditions with natural challenge inoculations by vectors. Bio/Technology *10* (1992) 1562-1570.
17. Oakes, J.V., Shewmaker, C.K., Stalker, D.M. Production of cyclodextrins, a novel carbohydrate, in the tubers of transgenic potato plants. Bio/Technology *9* (1991) 982-986.
18. Jongedijk, E., de Schutter, A.A.J.M., Stolte, T., van der Elzen, P.J.M., Cornelissen, B.J.C. Increased resistance to potato virus X and preservation of cultivar properties in transgenic potato under field conditions. Bio/Technology *10* (1992) 422-429.
19. Datta, S.K., Datta, K., Soltanifar, N., Donn, G., Potrykus, I. Herbicide-resistant Indica rice plants from IRRI breeding line IR72 after PEG mediated transformation of protoplasts. Plant Mol. Biol. *20* (1992) 619-629.
20. D'Halluin, K., Bossut, M., Bonne, E., Mazur, B., Leemans, J., Botterman, J. Transformation of sugarbeet (*Beta vulgaris* L.) and evaluation of herbicide resistance in transgenic plants. Bio/Technology *10* (1992) 309-314.
21. de la Fuente-Mart nez, J.M., Mosqueada-Cano, G., Alvarez-Morales, A., Herrera-Estrella, L. Expression of a bacterial phaseolotoxin-resistant ornithyl transcarbamylase in transgenic tobacco confers resistance to *Pseudomonas syringae* pv. *phaseolicola*. Bio/Technology *10* (1992) 905-909.
22. Smith, C.J.S., Watson, C.F., Ray, J., Bird, C.R., Morris, P.C., Schuch, W., Grierson, D. Antisense RNA inhibition of polygalacturonase gene expression in transgenic tomatoes. Nature *334* (1988) 724-726.
23. Vasil, V., Castillo, A.M., Fromm, M.E., Vasil, I.K. Herbicide resistant fertile transgenic wheat plants obtained by microprojectil bombardment of regenerable embryogenic callus. Bio/Technology *10* (1992) 667-674.
24. Jones D.A., Ryder M.H., Clare, B.G., Farrand, S.K. and Kerr, A. Construction of a tra^- deletion mutant of pAgK84 to safeguard the biological control of crown gall. Mol. Gen. Genet. *212* (1988) 207.
25. Ghosal, D., You, I.S., Chaterijee, D.K. and Chakrabarty, A.M. Microbial degradation of halogenated compounds. Science, *228* (1985) 135.
26. Karns, J.S., Kilbane, J.J., Duttagupta, S., and Chakrabarty, A.M. Metabolism of halophenols by 2,4,5-trichlorophenoxyacetic acid-degrading *Pseudomonas cepacia*. Appl. Environ. Microbiol. *46* (1983) 1176.

27. Bishop, D.H.L., Entwistle, P.F., Cameron, I.R., Allen, C.J. and Possee, R.D. Field trials of genetically-engineered Baculovirus insecticides. In: "Release of genetically-engineered microorganisms" M. Sussman et al., (Eds.). Academic Press, 1988, pp. 143–179.
28. USDA-APHIS. Environmental assessment and finding of no significant impact. Permit Number 90-333-01, 1991, pp. 1–37.
29. Ronson, C.W., Bosworth, A., Genova, M., Gudbrandsen, S., Hankinson, T., Kwiatkowski R., Ratcliffe, H., Robie, C., Sweeney, P., Szeto, W., Williams, M. and Zablotowicz, R. Field release of genetically engineered *Rhizobium meliloti* and *Bradyrhizobium japonicum* strains. In: Nitrogen Fixation: Achievements and Objectives" P. Gresshoff et al., (Eds.). Chapman and Hall Publ., 1990, pp. 397–403.
30. Lindow, S.E. Tests of specificity of competition around *Pseudomonas syringae* strains on plants using recombinant ice-strains and use of ice-nucleation genes as probes of *in situ* transcriptional activity. In: "Advances in Molecular Genetics of Plant-Microbe Interactions" vol. 1, H. Hennecke and D.P.S. Verma (Eds.). Kluwer Acad. Publ., 1991, pp. 457–464.
31. Birkenhead, K., Wang, Y.P., Noonan, B., Manian, S.S. and O'Gara, F. Characterization of *Rhizobium meliloti dct* genes and relationship between utilization of dicarboxylic acid and expression of nitrogen fixation genes. In: "Molecular Genetics of Plant-Microbe Interactions" R. Palacios and D.P. Verma (Eds.). Amer. Phytopathol. Soc., St. Paul Publ., 1990, pp. 834.
32. Huttner, S.L., Arntzen, C., Beachy, R., Breuning, G., Nester, E., Qualset, C., Vidaver, A. Revising oversight of genetically modified plants. Bio/Technology *10* (1992) 967–971.
33. OECD. Recombinant DNA Safety Considerations: Safety Considerations for Industrial, Agricultural and Environmental Applications of Organisms Derived by Recombinant DNA Techniques. OECD Publ. Serv., Paris, 1986.
34. USDA-APHIS. User's Guide for Introducing Genetically Engineered Plants and Microorganisms. USDA Technical Bulletin no. 1783, 1991.
35. Executive Office of the President, Office of Science and Technology Policy. Exercise of federal oversight within scope of statutory authority: planned introductions of biotechnology products into the environment. February 27, Federal Register *57* (1992) 6753–6762.
36. OECD. International Survey on Biotechnology: Use and Regulations. OECD Environments Monographs no. 39. OECD Publ. Service, Paris, 1990, pp. 1–113.
37. Commission of the European Communities. Methods for the detection of micro-organisms in the environment. Luxembourg, Office for Off. Publ. of the E.C., 1992, pp. 1–123.
38. Food and Drug Administration. May 26, 1992. Federal Register *57* 22984–23005.
39. Scarpa, M., Zacchello, F., Terapia genica delle malattie ereditarie. Prosp. in Pediatria *22* (1992) 247–255.
40. The Yeast Industry Platform. From tradition to high-tech: the yeast, products, prospects, impacts. YIP Secretariat, c/o Tech-know, av. de l'Observatoire 2, Brussels, 1992, pp. 1–12.

Index